From the Parade Child
to the King of Chaos

**COMPLICATED
CONVERSATION**

A Book Series of
Curriculum Studies

William F. Pinar
General Editor

Volume 49

The Complicated Conversation series is part of the Peter Lang Education list.
Every volume is peer reviewed and meets
the highest quality standards for content and production.

PETER LANG
New York • Bern • Frankfurt • Berlin
Brussels • Vienna • Oxford • Warsaw

Hongyu Wang

From the Parade Child to the King of Chaos

The Complex Journey of William Doll, Teacher Educator

PETER LANG
New York • Bern • Frankfurt • Berlin
Brussels • Vienna • Oxford • Warsaw

Library of Congress Cataloging-in-Publication Data

Names: Wang, Hongyu, author.
Title: From the parade child to the king of chaos:
the complex journey of William Doll, teacher educator / Hongyu Wang.
Description: New York: Peter Lang, 2016.
Series: Complicated conversation: a book series of
curriculum studies; vol. 49 | ISSN 1534-2816
Includes bibliographical references and index.
Identifiers: LCCN 2016023378 | ISBN 978-1-4331-3411-1 (hardcover: alk. paper)
ISBN 978-1-4331-3410-4 (paperback: alk. paper) | ISBN 978-1-4539-1905-7 (ebook pdf)
ISBN 978-1-4331-3597-2 (epub) | ISBN 978-1-4331-3719-8 (mobi)
Subjects: LCSH: Doll, William E. | Educators—United States—Biography.
Postmodernism and education. | Curriculum planning—Philosophy.
Chaotic behavior in systems.
Classification: LCC LB885.D65 W35 2016 | DDC 370.92 [B] —dc23
LC record available at https://lccn.loc.gov/2016023378

Bibliographic information published by **Die Deutsche Nationalbibliothek**.
Die Deutsche Nationalbibliothek lists this publication in the "Deutsche
Nationalbibliografie"; detailed bibliographic data are available
on the Internet at http://dnb.d-nb.de/.

The paper in this book meets the guidelines for permanence and durability
of the Committee on Production Guidelines for Book Longevity
of the Council of Library Resources.

© 2016 Peter Lang Publishing, Inc., New York
29 Broadway, 18th floor, New York, NY 10006
www.peterlang.com

All rights reserved.
Reprint or reproduction, even partially, in all forms such as microfilm,
xerography, microfiche, microcard, and offset strictly prohibited.

Printed in the United States of America

Dedicated to my nephew,

David (Fang) Man,

Who has shared precious moments of his childhood

and has been inspirational in his becoming

TABLE OF CONTENTS

	Introduction *William F. Pinar*	ix
	Preface: Complexity of Life, Thought, and Pedagogy	xxiii
Chapter 1.	Pedagogy of Play	1
Chapter 2.	Pedagogy of Perturbation	25
Chapter 3.	Pedagogy of Presence	45
Chapter 4.	Pedagogy of Patterns	69
Chapter 5.	Pedagogy of Passion	89
Chapter 6.	Pedagogy of Peace	111
Chapter 7.	Pedagogy of Participation	129
	Bill Doll's Pedagogy: A Peek Behind the Curtain *David Kirshner*	153
	Afterword: Of Experiencing Pedagogy by Rainbow Light *Molly Quinn*	167
	References	179

INTRODUCTION

William F. Pinar

It is no secret that I locate study, not teaching, at the site of educational experience.[1] Teaching matters, as the life and career of my dear friend and key colleague William E. Doll, Jr., make clear. One of the ways teaching matters is in its communication of subject matter—simultaneously the school subject and the human subject—the educator's engagement with students and colleagues in the complicated conversation that is the curriculum. At its consummate, teaching can be in consanguinity[2] with its subject matter. As Hongyu Wang observes, "Doll teaches what he is."[3] Wang acknowledges that "such a unity between the teacher and teaching is remarkable." Wang quotes the insight of another remarkable educator—Ted Aoki—"good teachers are more than they do; they are the teaching." So too is Hongyu Wang, whose teaching in this book is in consanguinity with the pedagogy of William E. Doll, Jr., a great American teacher educator and curriculum theorist.[4] It has been my privilege and pleasure to know Bill Doll for forty years, first in upstate New York, then in Louisiana, and now in the Pacific Northwest. Because I have before expressed my gratitude for this friendship[5] in personal terms, I will not repeat it here. Because I have composed an overview[6] of his resounding scholarship, I will not repeat that either. Here I focus on Doll's exceptional teaching and Hongyu Wang's wonderful teaching of it.

"Wonderful" seems precisely the word, as Wang works from "wonder," defined as both the desire to know something and feelings of admiration, amazement, and marvel. Severing teaching from the specificities of situations—conceiving teaching as technique or best practice—risks remaking education a matter of manipulation, communicating to children they are merely means to depersonalized ends: school test scores.

Partly to protect students, partly to protect themselves, policy makers have promoted a professionalism stripped of the personal, rendering personal relationship incidental and even suspicious, especially between educators and young children. Wang's achievement here is that she not only communicates, in consanguinity, the pedagogical genius of William E. Doll, Jr., she has also contributed to a conception of teaching that reintegrates what twentieth-century professionalism had split apart: subject matter and the human subject.

By emphasizing the seven modalities of Doll's teaching—play, perturbation, presence, pattern, passion, peace, and participation—Wang abstracts from the concrete without obscuring either. To invoke these poles of representation Wang testifies to her own experience studying with the great teacher. She also interviews others who have worked with Doll, allowing the particularity of those relationships, embedded in biographical and historical time and in certain places, to communicate the concept. Even the book's narrative itself she renders porous, ensuring that Doll and his students and their relationships reverberate throughout. Wang writes, "I don't see this as *my* book, even though I am the writer," but one "authored by all participants." Humility is also a trait of her teacher, as we will see. Note she uses tactile and auditory, not ocular, metaphors, when inviting readers to "listen to his teaching."[7]

A Parade of Pedagogies

Such consanguinity seems suitable, as Wang turns first[8] to play and the sound of Doll's laughter, still echoing in the ears of those who have been privileged to hear him. Like his teaching, Doll's "laughter is loud and long with a unique rhythm," as Wang describes. Play—simultaneously "intellectual, social, and spiritual"—pervades Doll's classroom, now not exactly a playroom, but still softened somehow, malleable as it accommodates twists and turns unanticipated by those who see such rooms as only serious and supine.[9] To supplement the point Wang again quotes Aoki, noting that work and play in our culture tend to be split. For Doll, the two are conjoined, providing a bridge—another famous image of Aoki's—to link what seems separate. There is a concrete referent for this abstract idea; Doll remembers a bridge on the Cornell University campus where he tossed his sunglasses, an event I would allegorically decode as discarding his blinders. He could see clearly now. Wang locates Doll's sense of teaching as a complex form of play in his upbringing, both as a child and, later, as a graduate student, commenting: "The combination of the two in contradiction, freedom to explore as a young child and as a graduate student, and confinement to set rules at school, paved the way for Doll to become a playful educator."

Playfulness is not only a personality characteristic, as Wang makes plain. It is for Doll an orientation to teaching; as Wang tells us, "he quickly learned that play, rather than seriousness, takes students further in their growth." That Deweyan idea seems serious to me—however playful its variable means of cultivation—a point Wang affirms, if indirectly, when she notes that "respecting students' humanness and their ability to surpass the teacher . . . became important in his teaching." That respect—affirming students' capacities to exceed what they have become, including what the teacher has achieved—becomes personified in the person of Donna Trueit, first Doll's student, then wife, then teacher and editor (see Trueit, 2012). Doll's conversations, Trueit tells us, had the consequence of "calling me out of myself." And Doll credits Trueit "with her amazing perceptions and phenomenal insights," with enabling him to more deeply appreciate difference.

"Playing with the limits of the system," Wang writes, "Doll encourages students to play with their own limits." She adds: "Playing with ideas is a distinctive mark of Doll's pedagogy in teacher education." Such play is "not only intellectual, but also aesthetic and spiritual." These dimensions of play became concrete as they were enacted within actually existing relationships: amidst ideas, between persons.[10] As complex as "play" is in Doll's theory and practice—with subtexts in Dewey, Morin, Piaget, Prigogine, Whitehead, and others—it inaugurates nothing less than a new curriculum paradigm:

> Starting with autobiographical experiences, running through a complex structure of curriculum as an open system, connecting with cosmological interdependence and interrelationship, and going through a transformative process of conversing with the text, the other, and the world, the new curriculum paradigm ties together the self and the communal, the human and the ecological, and the dynamic and the structural. At the same time, it is not a seamless whole, as creative tensionality is the driving force for generating new intelligence....

As Wang explains, in this paradigm Doll combines the "logical reasoning of science" with the "experiential" and the "relational" expressed in narrative with the "vitality of spirit." Through their "interplay" the curriculum—as studied, as taught, as lived—becomes open to a "sense of the sacred." Spirituality (not necessarily organized religion) has become increasingly important to Doll, Wang notes, emphasizing the significance of "humility" in the face of "our own finitude, our own flaws, and our own struggles."[11] Such humility implies that "we need to be able to move away from the individual self to society and, further, to the cosmos."

The Piagetian subtext noted earlier surfaces in Wang's naming of the second dimension of Doll's pedagogy: perturbation. Wang points to chaos and complexity theory as complementing Piagetian disequilibrium in theorizing "the right amount of tension" that is "beneficial for a system's transformation." (As you might imagine, that "right amount" includes a healthy dose of play.) Acknowledging that "confusions" and "errors" are often dreaded as undermining learning, Wang points out that in Doll's theory they are "key elements" for "initiating students' intellectual reorganization to reach a higher level." As did Aoki with his "both/and"[12] emphasis, Doll asks (in Wang's words) "educators to look beyond any either/or framework and welcome the roles of disequilibrium, perturbation, and uncertainty."[13]

The teacher must maintain a certain skepticism concerning one's own viewpoint in order to remain "open to students' fresh ideas." Such skepticism is not the province of educators' alone, as students too can profit by remaining in a certain disequilibrium, perhaps sometimes in the vicinity of but not nearby one's "breaking point."[14] Such vigilance can require constant contemplation of one's condition, as these states may not remain static, as they shift in intensity and scale according to circumstance, internal and external.[15] Eschewing "any predetermined content or specific objectives," Doll remained "open to unexpected twists and turns during the semester and [made] modifications and adjustments." I am reminded of Aoki's emphasis on improvisation;[16] in both instances educators must have some shifting sense—theoretical or intuitive or some combination of these two—about what knowledge is of most worth. Wang would seem to concur, asserting that Doll always had "a general sense of major ideas and frameworks in a class, while letting the specific details emerge with students' participation." One major idea was indeed perturbation. For herself, Wang invokes Kristeva's conception of intimate revolt[17] to affirm the "positive role"[18] perturbation can play in "recursive transformation."

However abstract these terms can be, they are enacted educationally within specific relationships: with and among children, with texts, colleagues, and others, living and deceased. That "right amount" of perturbation is, then, no matter of mathematics but of emotion and intuitive judgment, neither one of which is infallible. General guidelines are possible, however, as Wang notes: "relationships that support a pedagogy of perturbation must be both loving and challenging." Too much perturbation provokes "resistance or avoidance." Judgment is required to maintain the right amount of tension, including the right distance[19] between teacher and student.

The "right distance" between student and teacher is a matter of subjective presence threaded pedagogically.[20] In Chapter 3, Wang details the forms Doll's pedagogical presence took, among them sharing food and wine as well as ideas. "Sharing" would seem to be the salient word, as through various forms Doll taught not only who he was but what an educator can be. Such teaching represents subjective presence within relationships and situations, as Wang emphasizes: "William Doll's presence as a person who is genuinely interested in students as persons created a pedagogical bonding that remained unique and powerful for many of his students."

What proved "unique" and "powerful" to Doll himself, Wang reports, followed from his team teaching experiences with Hendry, Trueit, and Fleener,[21] as these left him "more conscious of [his] . . . male control."[22] As Wang observes: "To be present to oneself, one needs the help of others. In this sense, a pedagogy of presence means not only to be present to students but also to be present to the self, and often the two aspects are intertwined." Like the dyadic effect, presence can proceed through reciprocity.[23] In Doll's pedagogy of presence, relationality resides "at the heart of teaching and learning."[24] Guiding students' intellectual, emotional, and spiritual transformation, Doll provided "pedagogical companionship" and "sustained engagement," thereby contributing to "students' own sense-making and meaning-making." Wang confides: "He is still that guide for me."

Given the primacy of relationship in teaching, that final preposition should be underlined, which Wang does when she concludes: "[H]e is present personally with students in their adventures, intellectually and socially." Congruent with his emphasis on the self in the system, Doll sees not only the system but also persons when he teaches. He acknowledges each person's individuality. There may be no individuality without relationship, but it bears mentioning that there is no relationality without individuality. In classroom cultures of conformity and compulsory collaboration, both are threatened.

Doll's presence in his teaching is also expressed through his remembrance of what students have said, weaving what has been said before to what is being said now. "He remembers each student's interest," Wang recalls, and during class he addresses it, sometimes referencing "additional resources," at other times "other students' comments." This would occur not only within one class, but also over the course[25] of the class, returning to "a student's comment a week or a month later to build further connections." Invoking one of Doll's 4 R's (see Doll, 1993, p. 156), Wang concludes: "A recursive time situated in the present invites students to become participants."

Through "presence," Wang writes, Doll engaged "difference," as when women colleagues reminded him of his gendered tendencies toward "control." Doll made modifications in his thought and conduct. Race too had been at first a "blind spot," Wang recalls, leaving several of his (including minority) students "skeptical," but Doll did not withdraw: "his pedagogical commitment to these students was strong." Race, class, and gender are not the only markers of difference, of course; there are differences of style, informed as these may be by the former. Wang forefronts differences between Trueit and Doll irreducible to those, such as "Doll's spontaneity," which, Wang tells us, "clashed with Trueit's preference for organization." Another difference between the two teachers is that Trueit was "more attuned" to "negative" students' perceptions, while Doll seemed more attuned to "the positive side." Doll's pedagogical "presence" includes both being "absorbed into an experience and acting upon the world." In my terms, subjective reconstruction and social reconstruction are reciprocally related.

The poles of self and system resurface in Wang's discussion in Chapter 4 of the roles "patterned relationships" play in Doll's pedagogy. These are not restricted to the intersubjective sphere, as Wang lists four senses of the term embedded in Doll's "pedagogy of patterns." The first sense is the patterning discernible in each subject; the second is the patterning evident in "students' thinking." Bridging the patterns of a subject and a student's thinking, Wang reminds, is "play." Third would seem to be the shifting patterns of "patterning" itself—"emergent" and "nonlinear"—meaning that a "pedagogy of patterns forms and transforms itself." Fourth, patterning is "both ecological and spiritual," linking the human with "all living creatures," a fact that (again in Wang's words) "leads to awe, wonder, and humility toward the interconnectedness of life that is ever shifting." Wang summarizes: "Doll's pedagogy of patterns emerges from a nonlinear patterning of his life, thought, and teaching, which is both recursive and open." That patterning, Wang adds, is animated by "passion."

William Doll is "passionate" about teaching, Wang writes, and so (again in consanguinity) he teaches "passionately." In Chapter 5, Wang discusses Doll's pedagogy of passion, expressed through "his joyful spirit as a person, his devotion to students as a teacher educator, his enthusiastic play with ideas as a scholar, and his eagerness to learn from others in a distant country."[26] Despite his Catholic faith, Doll does not enact the suffering associated with the Passion of Christ. Secular traces of that faith remain, evident when Wang writes that his work with students intends them "to move upward," as he encourages their "search for meaning" in states of open-mindedness, always "willing to take intellectual risks." Surely that is a pedagogical creed of faith, a commitment to transcendence through learning, a time-honored tradition in Catholic education.[27] Consistent with Catholicism's emphasis on the Passion, Doll locates "passion at the heart of teaching and learning." Also consistent is Wang's characterization of his pedagogy of passion as "an embodied way of teaching." No Last Supper, Friday Friends met at lunch: "sharing food and sharing ideas are inseparable for Doll," Wang notes. Conversation, and not only over lunch, was, Wang writes, also embodied, as students convulsed in "laughter" even as they remained poised in attentiveness to others. "Sometimes he closes his eyes while listening, but his ears are open and receptive, and when he opens his eyes, he always has something intriguing to say," she remembers.[28] From her study of Doll's pedagogy of passion Wang concludes: "Students' curiosity, intuition, and emotional investment in learning can be activated by a teacher's joyful commitment to them." Suffering may remain silent in his pedagogy of passion, but Doll's self-sacrifice to his teaching is constant and clear.

Wang follows her discussion of a pedagogy of passion with what she terms Doll's pedagogy of peace, itself "dynamic and closely related to his sense of play." In Doll's pedagogy, Wang suggests, "peace keeps passion balanced and passion keeps peace alive." Conflict, for instance, requires "firmness" and "calmness": these are, she continues, "the double gestures of nonviolent engagement." Relationality[29] is paramount: "Playing with the limits and playing with social relationships would not be possible without a certain flow that springs from inner peace." Doll's pedagogy of peace "does not teach students directly; instead, students experience peace through their pedagogical relationships with the teacher to enable learning."

Wang ascribes Doll's sharp sense of peace to his childhood and specifically to the loss of his father. The theological associations of passion surface again when Wang writes: "Therefore, his way of making peace with loss [Wang is alluding to his father's death] and pain [including physical pain] has shaped an outlook that focuses on moving forward by engaging in meaningful activities." Doll's joy is no denial of suffering but, one might say, its transubstantiation. Perhaps something remains of Catholic guilt as, Wang writes, "when a student is not successful, it weighs heavily on him, and he keeps wondering what he could have done differently to help the student."[30] Doll's distress over each student's fate demonstrates that simple fact. Guilt can be good and not necessarily tormenting: Wang notes that Doll is "at peace with himself as a human being with blind spots." Such "an active form of peace," she adds, "becomes possible only through working with discomfort, difficulty, and difference."

In Doll's pedagogy of peace—Wang emphasizes "nonviolence" (see Wang, 2014a, 2014b)—"teaching in which engagement is not instrumental but existential." In Doll's pedagogy "flow, peace, and nonviolence come together through engaging difference."[31] Not only the educator experiences such synthesis, students can as well; Wang observes that a pedagogy of peace has "an integrative effect on students." She suggests that "Doll's own journey of making peace with trauma, loss, and guilt in his life, his patience with students who go through various difficulties, and his compassion for those who suffer make it possible for him to connect with students and bring them a certain sense of peace." Wang summarizes: "Enacting emergent control, forming peaceful relationships with the self and the world, accepting the positive role of tensions, and creating collaborative teaching relationships all contribute to Doll's pedagogy of peace."

In Chapter 7, Wang describes Doll's pedagogy of participation, his capacity for building communities in which "relationships are not only between the teacher and students but also among students." This pedagogy of participation is "not so much about using any particular techniques," but about Doll's "capacity to connect with others and help others to connect." In Doll's classes, Wang adds, "everybody participates."

Doll's pedagogy of participation extends beyond the classroom to "local, national, and international communities." His capacity for building communities across borders has been conspicuous. It was Doll's idea to host at Louisiana State University (LSU)—where we worked together for almost twenty years—international conferences in 1999 and 2000. The former coincided with an international philosophy of education meeting held in New Orleans, and for the latter—Donna Trueit's participation proved pivotal, as was Hongyu Wang's—we invited colleagues worldwide to engage in internationalizing their fields and ours.[32] What underlies this conspicuous commitment to community? It is, Wang suggests, "a strong sense of responsibility—political, ethical, and ultimately spiritual responsibility."

The sense of participation Doll wants for his prospective and practicing teachers extends, then, beyond professionalism narrowly stipulated to a civic sense of commitment to teachers' own local, regional, and national communities. For Doll, Wang explains, "participation leads to the transformation of the participants and the community through interactions." One way he modeled this interaction for students was through team teaching, in which "the instructors' respect for each other, their willingness to de-center authority roles in the classroom, and their ability to negotiate and play with difference demonstrate[d] for students what collaborative participation and mutual learning can be." Wang points repeatedly to the "affirmative presences" of both Petra Hendry and Donna Trueit who "gently challenged" him to be more attentive to questions of the Other. Al Alcazar noticed that Doll's language "became inclusive over time." Wang concludes: "Doll's philosophy of interactive, transformative, and democratic participation that leads to transformation has worked magic for his own change as well."[33] Such subjective reconstruction invites social reconstruction, as Wang implies:

> If letting things emerge in complex class dynamics becomes a form of politics that replaces the power struggle with everyone's participation, then the political becomes ethical, which shifts the relational dynamic from the win-or-lose mentality to the share-and-transform mentality.

Here Wang is importantly pointing to what one hopes is the next moment in U.S. curriculum studies, wherein the primacy of the political is replaced by an ongoing engagement with the ethical.[34]

Engagement with the ethical invites studies in spirituality, also implied when Wang writes that Doll's strong sense of "responsibility" occupies "the intersections of politics, ethics, and spirituality," linking the three orders of interpretation and understanding in "a complicated space in which each pathway is simultaneously contested and enriched." Just as the "dynamics of participation cannot remain static but must be open to uncertainty and incompleteness," she adds, "the political, the ethical, and the spiritual are all embedded in relationships that emerge from interactions." Such interactions include texts, I add, sometimes studied in solitude.

Conclusion

William Doll's pedagogy of participation, Wang points out, is interlaced with his pedagogy of play, perturbation, presence, patterns, passion, and peace, the seven forming together "a complex web of teaching, life, and thought." That "web" also constitutes a form of "self-organization," what Wang terms Doll's "central curriculum metaphor," one that "works within and across his life, thought, and pedagogy to bring forth the magic of his teaching." The "centering" of any "individual entity"—whether that entity is an "individual person, or a text, or society"—cannot "capture the complexity of education." I appreciate, as Wang explains, that it is the ongoing "interaction among all components in the system that fertilizes moments of self-organization," but is it not also the individual person who "fertilizes" a sometimes seemingly sterile system, a fact to which Doll's own pedagogy so powerfully attests?

It has been Doll's own indelible distinctiveness from which he has worked to encourage the distinctiveness of his students. Doll's "pedagogical concerns," Wang summarizes, focus on "how students can develop their own teaching approaches that may or may not mirror his own." That encouragement of pedagogical difference, of becoming one's own person—one's own teacher, distinctive and determined to learn from one's experience—while remaining committed to others may be what I have admired most of all while working alongside Bill Doll. Despite the stress of "systems" and the spinning of "webs" the King of Chaos is very much a New England individualist.[35] And he has, over the decades at different institutions and in different ways, invited his students to become individualists as well.

"Doll's passionate teaching sparks and informs students' learning through the network effect," Wang emphasizes, "but does not center on the teaching self." Perhaps—the emphasis upon "self-organization" surely subsumes within it the teaching self—but I cannot help but notice that this book is about one quite exceptional "teaching self," a book composed by a second exceptional self, one in complicated conversation with herself, with Doll, with his students, and with readers who know none of us. Doll may "not [be] much concerned with the question of subjectivity,"[36] but, as Wang observes, "students' interior world is touched by his teaching, so a space for subjective and intersubjective transformation is opened up." I do not doubt, as Trueit points out, that "Bill's notion of the self is the self-in-the-system." But it seems also true, as Wang notes, that "to create a space for personal growth and transformation" one "works on oneself through working with the system."

Like the seven modalities of Doll's pedagogy Wang has so skillfully described, "self" and "system" are inextricably intertwined. But they are also separable. This fundamental fact is indicated in Wang's depiction of Doll's "encounter with China," which provoked, she reports, "a process of self-education through his passionate learning from the other." This "turning inward" meant for Doll a questioning of the self—specifically he questioned his own privilege—one that indicated, Wang notes, the "mutuality in his engagement with the other."[37] Certainly that has been my experience with William E. Doll, Jr. An older brother, dear friend, and fellow teacher, to me Bill Doll is indeed a "King" and not only of chaos.

NOTES

1. Pinar, 2015, pp. 11–24.
2. Francine Shuchat Shaw (1975/2000, p. 445) proposed this concept forty years ago.
3. Unless otherwise indicated, all quoted passages come from this book.
4. Anne Phelan (2015) has demonstrated definitively the reciprocity between the two domains. Wang's study emphasizes Doll's significance for the field of teacher education.
5. Pinar, 2006, pp. xiv–xv.
6. Pinar, 2012.
7. Ted Aoki emphasized the auditory in calling for curriculum in a new key. See his collected works (Pinar & Irwin, 2005).
8. Wang insists the book, like Doll's teaching, is "nonlinear" but I appreciate her sequencing, as surely "play" is the king of this parade of pedagogies. "It seems quite fitting," Wang admits, "to open this book with his pedagogy of play, an element *essential* to his life, thought, and teaching" (emphasis added).
9. As in "inactive, passive, or inert, especially from indolence or indifference," defined by dictionary.com, retrieved March 8, 2016.
10. As Wang notes, play was not limited to the classroom, but occurred also over food and drink, regularly in meetings of the Friday Friends and in parties at the Doll-Trueit home in New Orleans and at the Pinar-Turner home in Baton Rouge. Students and faculty met informally during such occasions that also featured distinguished guests, for example, Michel Serres. The party Doll and I hosted at my apartment (after my son returned to California, before I met my husband) on Esplanade in New Orleans during the 1994 AERA meeting

extended Doll's sense of play to colleagues; among the fifty or so guests were Maxine Greene, Petra Hendry, Joe Kincheloe, Peter McLaren, and Shirley Steinberg.
11. "Humility," Wang explains, "is an important element of Doll's sense of spirituality."
12. See Pinar and Irwin, 2005, pp. 260, 271, 282.
13. Wang references Doll's experience with Steve Mann (1975/2000), who challenged Doll to "unlearn" the conceptions of citizenship he had acquired as a child to incorporate dissent, even civil disobedience.
14. Wang quotes Petra Munro Hendry (2011)—the important curriculum theorist and historian—who, after teaching several classes with Doll, remembered that he would sometimes pose "the most outrageous or bizarre questions just in order to shake people up and get them thinking." Wang adds: "Maintaining a dynamic tension 'between challenge and comfort' (Doll, 1989/2012, p. 203), he has learned to guide students to take steps in venturing to the uncertain yet still feeling connected."
15. In a passage that Wang quotes, Doll makes this point: "Because no pre-set formula can tell a teacher what this [just the right amount of chaos to be creative] will be for individual students, teaching becomes an art."
16. See Pinar & Irwin, 2005, p. 367.
17. See Wang, 2004, 2010.
18. That positive role seems to be precipitation (or initiation), evident in Wang's choice of gerunds: "Doll . . . has developed the connection between Piaget's work and Prigogine's work, in which perturbation plays a triggering role in intellectual development and personal growth."
19. See Taubman, 1990/2015.
20. "Presence," Wexler (1996, p. 144) writes, "is existential, concrete, mutual and inclusive of the other in a way that goes beyond psychological empathy."
21. See Doll et al., 2005.
22. Each of us is a creature of our generation; education enables each of us to reconstruct what time and place and parents have made of us. For Doll, his gender formation also had its welcome elements, as Wang notes when she quotes Stephen Triche's characterization of him as a "paternalistic postmodernist." According to Wang, "the combination of parental devotion to the younger generation and the fluidity to allow freedom in postmodernism is actually a good match for Doll's style of pedagogical engagement." That engagement contains a sense of "responsibility," as Wang also notes: "Doll's sense of responsibility is tied to his sense of presence. One must be there to meet the other and respond to the other."
23. Wang writes: "As a result of his enthusiasm, students have also become more present to one another."
24. "Wholeness is not within the self but in relationships," Wang writes. For me, there can be no either/or here, as the self–self relationship informs, as it is informed by, the presence of others.
25. And outside class as well, as Wang writes: "Even without a physical presence, Doll is still in contact with students through emails, phone calls, and letters to keep conversations going." Later Wang registers that "Doll not only engages with students in the class but also maintains his presence outside of the classroom and after students' graduation."

26. While he taught in many countries, it was in China—Wang accompanied him and Trueit on an extended lecture tour there in 2000—where Wang watched "Doll's passion for learning *from* the other." Wang adds: "His keen observations gave me a fresh lens through which to reexamine my own culture."
27. See Petitfils, 2015, pp. 14–43.
28. Donna Trueit told Wang that "intellectual improvisation" has always been one of Doll's strengths, a concept associated with another great teacher: Ted Aoki.
29. Referencing Jane Addams, Wang affirms that "ideals of peace are not tranquil or passive but complex and dynamic, leading to the nurturance of public life."
30. After a lifetime of ambivalence toward it, finally I have come to agree with Lasch that guilt can be good. Conscience, he writes, originates not so much in the "fear of God" as in the "surge to make amends" (1984, p. 259).
31. Perhaps such teaching is both existential and instrumental. A gentle linking of ends and means occurs in Wang's depiction of Doll's devaluing of knowledge—another one of our cordial disagreements—when she notes (beautifully, it seems to me) that "being in its flow is at the center of his pedagogy." She adds: "It is not that the standard is not important for Doll—he has worked hard to improve the rigor of the teacher education programs—but he is skeptical of any predetermined standard imposed from outside." Isn't the demand that nothing be predetermined also a standard imposed from outside?
32. As I learned in my studies of curriculum studies in five countries, "their" fields were already internationalized; it is the U.S. field that has been self-enclosed. Several of the papers presented at the 2000 LSU Conference on the Internationalization of Curriculum Studies were collected and published: see Trueit et al., 2003.
33. In Wang's work with multicultural issues, such "subjective reconstruction" figures prominently, as she suggests that students and teachers must sometimes "unlearn" racialized and gendered assumptions about others. "Unlearning involves the capacity to step out of oneself," Wang (2009, p. xiv) writes, "to look at the self from a distance in a new light."
34. Certainly it is my hope (Pinar, 2013, p. 76).
35. As an individualist, Doll worked to avoid "associating with any particular camp [within the field of curriculum studies], and he does not approach such non-belonging in a negative way." Indeed, Wang tells us that "Doll welcomes being 'out of place,'" an experience chronicled brilliantly—and individualistically—by Wang herself: see Wang, 2004.
36. "Not paying close attention to subjectivity is also related to Doll's ability not to take himself too seriously," Wang suggests. Since teaching is an expression of himself—as the Aoki insight earlier attests—it seems to me Doll is obligated to take himself seriously. Certainly he takes his students seriously. Care for self and care for others seem reciprocally related (see Jung, 2016, and Moghtader, 2015).
37. "A pedagogy of passion through ethical engagement with difference requires critical self-reflection," Wang writes, a pedagogy she demonstrates powerfully in her 2004 book.

REFERENCES

Doll, William E., Jr., Fleener, M. Jayne, Trueit, Donna, & St. Julien, John. (Eds.). (2005). *Chaos, complexity, curriculum, and culture.* New York: Peter Lang.
Doll, William E., Jr. (1993). *A Post-modern perspective on curriculum.* New York: Teachers College Press.
Hendry, Petra Munro. (2011). *Engendering curriculum history.* New York: Routledge.
Jung, Jung-Hoon. (2016). *The concept of care in curriculum studies.* New York: Routledge.
Lasch, Christopher. (1984). *The minimal self: Psychic survival in troubled times.* New York: Norton.
Mann, John Steven. (2000). On student rights. In William F. Pinar (Ed.), *Curriculum studies: The reconceptualization* (pp. 167–172). Troy, NY: Educator's International Press. (Original work published 1975)
Moghtader, Bruce. (2015). *Foucault and educational ethics.* New York: Palgrave Macmillan.
Petitfils, Brad. (2015). *Parallels and responses to curricular innovation: The possibilities of posthumanistic education.* New York: Routledge.
Phelan, Anne. (2015). *Curriculum theorizing and teacher education: Complicating conjunctions.* New York: Routledge.
Pinar, William F. (2006). *The synoptic text today and other essays. Curriculum development after the reconceptualization.* New York: Peter Lang.
———. (2012). Introduction. In Donna Trueit (Ed.), *Pragmatism, postmodernism, complexity theory: The fascinating imaginative realm of William E. Doll, Jr.* (pp. 1–10). New York: Routledge.

———. (2013). *Curriculum studies in the United States: Intellectual histories, present circumstances.* New York: Palgrave Macmillan.

———. (2015). *Educational experience as lived: Knowledge, history, alterity.* New York: Routledge.

Pinar, William F., & Irwin, Rita L. (Eds.). (2005). *Curriculum in a new key: The collected works of Ted T. Aoki.* Mahwah, NJ: Lawrence Erlbaum.

Shaw, Francine Shuchat. (2000). Congruence. In William F. Pinar (Ed.), *Curriculum studies: The reconceptualization* (pp. 445–452). Troy, NY: Educator's International Press. (Original work published 1975)

Taubman, Peter M. (2015). Achieving the right distance. In William F. Pinar & William M. Reynolds (Eds.), *Understanding curriculum as phenomenological and deconstructed text* (pp. 216–233). Kingston, NY: Educator's International Press. (Reprinted from *Educational Theory*, 40(1), 121–133, 1990)

Trueit, Donna. (Ed.). (2012). *Pragmatism, post-modernism, complexity theory: The fascinating imaginative realm of William E. Doll, Jr.* New York: Routledge.

Trueit, Donna, Wang, Hongyu, Doll, William E., Jr., & Pinar, William F. (Eds.). (2003). *The internationalization of curriculum studies.* New York: Peter Lang.

Wang. Hongyu. (2004). *The call from a stranger on the journey home.* New York: Peter Lang.

———. (2009). Introduction. In Hongyu Wang & Nadine Olson (Eds.), *A journey to unlearn and learn in multicultural education* (pp. xi–xxi). New York: Peter Lang.

———. (2010). Intimate revolt and third possibilities: Co-creating a creative curriculum. In Erik Malewski (Ed.), *Curriculum studies handbook: The next moment* (pp. 374–386). New York: Routledge.

———. (2014a). A nonviolent perspective on internationalizing curriculum studies. In William F. Pinar (Ed.), *International handbook of curriculum research* (2nd ed., pp. 67–76). New York: Routledge.

———. (2014b). *Nonviolence and education: Cross-cultural pathways.* New York: Routledge.

Wexler, Philip. (1996). *Holy sparks.* New York: St. Martin's.

PREFACE

Complexity of Life, Thought, and Pedagogy

> Shifting one's attitude from "reducing" complexity to "embracing" what is always already present in relations and interactions may lead to thinking complexly, abiding happily with mystery. (Doll & Trueit, in Doll, 2010/2012, p. 172)

> It is life that is characterized, above all, by transformation. Informed by evolution, history, politics, religion, Doll's very human life is dynamic, emergent, self-organizing, relational. (Pinar, in Doll, 2012, p. 8)

While this book is about an extraordinary teacher and teacher educator's complex journey, the formation of this book is itself complex. Writing it has taken much longer than I ever imagined, as it has gradually added layers. When I started this project, I thought this book would be published in time to celebrate the eightieth birthday of William E. Doll, Jr.,[1] but now he is approaching his eighty-fifth birthday! The difficulties have been multiple and my initial idea was reformulated several times to enable what has emerged.

I have learned from outstanding teachers and teacher educators both in China and in the United States, but William Doll is simply exceptional in his ability to reach out to students, create a class community, and play with ideas in the classroom. When I was a doctoral student at Louisiana State University (LSU) in the late 1990s, as far as I know every student enjoyed his teaching.

I often wondered what had contributed to his magical ability to teach. My initial questions were: What life experiences contributed to the formation of such a teacher educator? What school teaching experiences contributed to his pedagogy in teaching and mentoring pre-service and in-service teachers? How has the process of his subjective formation and re-formation as an educator unfolded?

As it turns out, this book addresses the first two questions: life experiences and school teaching experiences in their relationship with his pedagogy in teacher education. Although issues related to subjectivity are central to my interest, I quickly found out through interviews that Doll was not much concerned with subjectivity. For him, the notion of the self is always found in the situation and the system, rather than as a separate entity to be reflected upon. Certainly subjectivity does not simply disappear but works in a different way, is more dispersed, or, as Doll would say, "dissipative."

Another shift from my original plan is replacing "postmodern" with "complexity" in the book title. When I came to LSU in 1996, Doll's internationally acclaimed book, *A Post-Modern Perspective on Curriculum* (1993), was still fairly new, so there was a lot of attention paid to his postmodern approach. It did not take me long to realize that his version of postmodernism was quite different from other, more popular versions, but he had been known in the field of curriculum studies as a postmodernist scholar and known by his students as a "happy postmodernist," although joy and postmodernism did not seem to go hand in hand. I became curious and wanted to probe deeper into the formation of a postmodern educator: What was at work to lead him on this particular out-of-the-ordinary path? The original subtitle of the book was *The Formation of a Postmodern Educator*.

The Complex Journey is Doll's own preference and I made the change accordingly. As the collection of his work indicates (Doll, 2012), both postmodernism and complexity theory have played predominant roles in his scholarship and teaching, but his interest in complexity theory has grown along with his concern about its implications in education. I think complexity is a better metaphor, as it depicts Doll's interdisciplinary intellectual adventure and multilayered weaving of different strands to inform his teaching. Thus complexity rather than postmodernism becomes the central thread of this book. However, there is no linear progression from postmodernism to complexity; they are intertwined in Doll's thinking and teaching, and he makes an explicit connection between the two.

I chose a topical life history approach (Goodson & Sikes, 2001; Munro, 1998; Wang, 2014) because it was the journey behind William Doll's pedagogy that really interested me. Although there is a strong personal motivation for writing this book, it also fits with the needs of the large field of teacher education. The research on pedagogy of how teacher candidates are taught and how specific teaching approaches influence their learning has been extensive. However, there is little research to understand the relationship between teacher educators' life experiences and their pedagogical orientations. The focus on teaching methods and techniques can hardly reveal the richness and complexity of teaching (pre-service and in-service) teachers who encounter the complexity of teaching in their own work.

In recent decades a body of research on teachers' life stories and teachers' lore has explored how teachers' lived experiences influence their teaching and their identity as teachers (Goodson, 1992; Li, 2002; Schubert & Ayers, 1991). But little has been done on teacher educators' journeys through life, thought, and pedagogy. The report of the AERA Panel on Research and Teacher Education (Cochran-Smith & Zeichner, 2005), and two editions of *The Handbook of Research on Teacher Education* (Houston, 1990; Sikula, 1996) reveal that we have limited knowledge of teacher educators, and life history studies of teacher educators are specifically recommended for further inquiry.

This book addresses this need by using a life history approach to understand how one teacher educator's pedagogical orientations are related to his life experiences situated in the historical, social, and cultural context and related to American teacher education as a field. William Doll is an internationally renowned curriculum scholar who has worked in teacher education for decades. As his colleague and friend William Pinar points out, teaching and scholarship on teaching have been an essential concern of Doll's throughout his career as an educator (interview with Pinar, 2012). Studying Doll's half century of pedagogical life history offers many lessons for us.

Story-Sharing and Analysis

The term "data collection" seems inauthentic, as my interviews with William Doll's colleagues and former students were more the mutual sharing of stories in an effort to understand Doll's life, thought, and pedagogy more deeply. Interviews with Doll felt like listening to a wise elder sharing his stories, insights, and wisdom with the younger generation. The rich materials that I have collected are not "data" for me; instead, I use the term "story-sharing." In the story-sharing process, I interviewed Doll in the summers of 2009 and 2010, formally and informally interviewed his students and colleagues for the past seven years in both the United States and Canada, obtained the videotape of a class he and Donna Trueit co-taught at the University of British Columbia (UBC) in 2013, and throughout the process collected documents that were related to Doll's teaching.

First, interviews and email sources are major components of this study. My interviews with Doll during the summers of 2009 and 2010 took place in his home in Victoria, and gave me about thirty hours of tape-recorded conversation. The informal conversations with him and Trueit (his wife and team teaching partner for multiple classes) during those periods were also recorded. I also interviewed Donna Trueit for several hours in the summer of 2010. In addition, during the summer of 2009, I interviewed an international student from Russia, Anastasia Chebakova, who had taken a class with Doll and Trueit at the University of Victoria.

In December 2015, I communicated with Mary Aswell Doll (Doll's first wife) through email and she added valuable insights into William Doll's life and teaching. I interviewed two of Doll's former students, Al Alcazar and Stephen Triche, in Louisiana in 2012. Former students Mei Wu Hoyt and Jie Yu, both originally from China, communicated with me through emails and wrote short essays in 2010, from which I quote as personal communication. I also recently emailed another former student, Tayari Kwa Salaam, with a particular question related to Doll's teaching and African American students, and she responded through email. Sheryl Waltman, a former student in the Holmes program at LSU, also generously corresponded with me through email.

Doll's colleagues whom I interviewed include Petra Munro Hendry in 2011 at the Bergamo conference and William Pinar in 2012 when he visited Oklahoma State University. While attending a conference in Toronto, I took the opportunity to interview Wendy Kohli and later recorded notes from our conversation. I invited David Kirshner, a professor in mathematics education at LSU, to discuss his experience of Doll's teaching, and he graciously wrote a chapter on it, located at the end of the book, long before I finished this manuscript. I transcribed all the interviews and received permission from all interviewees to use the quotes from them.

Second, with deep thanks to Jung-Hoon Jung's hard work in helping coordinate everything, I obtained the videotape of a class that Doll and Trueit team taught in the spring of 2013 at UBC. The class members allowed Jung-Hoon to videotape the class, which I transcribed. Parts of the transcription are included as teaching episodes. Students whose conversations are included in the book gave me permission to use them. These episodes add a significant layer to demonstrate the actual process of Doll's teaching.

Third, Doll's publications and writings, course syllabi, letters, and other documents related to his teaching have been collected. Bringing together all these sources, I discuss Doll's pedagogy situated in the historical context of teacher education and the intellectual heritage of curriculum studies. In understanding his stories in personal, cultural, and educational contexts, I also bring in diverse literature to analysis.

As I probe the stories, interviews, documents, and teaching videos generated from this study, I come up with seven themes to present his complex journey. To follow Doll's postmodern playfulness with words, I call them the pedagogy of play, pedagogy of perturbation, pedagogy of presence, pedagogy of pattern, pedagogy of passion, pedagogy of peace, and pedagogy of participation. These seven P's, presented in Chapters 1 through 7, are certainly intertwined threads that cannot be neatly separated, and the complexity of interplay among these threads leads to the dynamic web of Doll's pedagogical life history.

When writing these chapters, I have followed a nonlinear path, sometimes writing chapters simultaneously rather than in a linear sequence, because everything is connected. Each theme presents a fractal entrance into a circle and exit out toward another circle. Since nonlinear teaching is a landmark of Doll's complex pedagogy, my writing about it is not linear. In these chapters, multilayered, recursive conversations are presented in a mixture of writing genres. Conventional academic writings are blended with stories, episodes from teaching videos, and my autobiographical writings (which are in italics). Several teaching scenarios are also reconstructed from interviews with colleagues who team taught with him, to create a humorous effect. In this way, the historical moments of his life, thought, and teaching are juxtaposed with his current thought and teaching, others' perceptions of his teaching are presented along with his own teaching stories, and my own stories add an intimate touch. The various storylines might not be compatible, but the point of complexity is to hold all layers together through dynamic interplay.

A Brief Sketch of William Doll's Teaching Career

Because Doll's stories in the next seven chapters do not follow a chronological order, here I provide a very brief sketch of his teaching career for readers as the context for thematic weaving.

William Doll was born in Detroit in 1931. His family lived in a small town near Boston when he was young. His father passed away when he was twelve years old, after which his mother single-parented him. As a schoolboy he had extensive experience working part-time. His parents allowed him to explore the world as a young child but he felt constrained in schools, even though he did not have problems academically. His teenage years were important in shaping his particular pathways in life. He attended Cornell University for his undergraduate studies, like his father, majoring in mathematics and philosophy. Graduating from college in 1953, he was not sure what his career path would be.

As it happened, he became a teacher without any formal training. His first teaching job was in New Hampshire in 1954, but he stayed for only one semester. In the fall semester of 1954, he took a job at Park Elementary School in Brookline, Massachusetts, and taught mathematics there for four years. During this time he began his master's level study, writing a thesis on Plato's theory of education. He received his master's degree from Boston University in 1960.

Before receiving his master's degree, in the fall semester of 1958, Doll went to Denver and became the Director of Mathematics at the Graland Country Day School, which served grades one through nine. He stayed there until 1963. During these years of teaching mathematics, he developed a series of strategies and devices—some quite unconventional—for teaching schoolchildren to play with the patterns of mathematics. He also participated in the Great Books program one year and had lively conversations with middle school students about literature. The ideas that he developed from his teaching experiences became the basis for his further inquiries into pedagogy and curriculum theory.

In the fall semester of 1963, he became the headmaster at Valley School in Baltimore. He encouraged teachers to understand their students by trying different teaching strategies without using teachers' manuals. He also played a leadership role in creating a school community in which parents, local artists, and community leaders were all involved in school activities, and connected with a teacher education program in a nearby college to have student interns at his school.

Doll started his doctoral studies at the Johns Hopkins University in the fall semester of 1967 as a full-time student. He finished his program in 1971 with a dissertation on John Dewey's educational thought. He appreciated the intellectual freedom and interdisciplinary orientation in his doctoral studies, and particularly his mentor, Steve Mann, who had a profound influence on him. Doll experienced a turning point in his life during that period; it was also a period of dramatic social change in the United States.

In the fall semester of 1971, Doll joined the faculty at the State University of New York (SUNY) at Oswego and worked there until 1983. Teaching courses in teacher education, he was not only interested in curriculum change in the College of Education but also led a university-wide curriculum change in the structure of general education. His ability to handle perturbations creatively certainly grew during these experiences. In 1983, he went to the University of Redlands as the Director of Teacher Education and worked there until 1988.

In 1988, Doll moved to Louisiana State University to join William Pinar, retiring in 2007. Those years of working at LSU were the richest for Doll, both in his scholarship and in implementing his curriculum vision, which he had not been able to accomplish fully at other universities. He particularly takes pride in LSU's elementary education Holmes program, which he designed, and the Curriculum Theory Project that he and William Pinar established to create a curriculum community. Many stories in this book come from that period.

After retiring, Doll moved to Victoria, Canada. He still teaches occasionally at the University of Victoria as an adjunct faculty, together with Trueit, using the format of team teaching. He joined William Pinar at the University of British Columbia and also team taught with Trueit there when possible, mainly because his passion for teaching has not faded with retirement. Although teaching as a career was serendipitous for him, it has been a perfect match. Doll has been an inspiration to students and teachers. His lifetime devotion to teaching and scholarship in pedagogy and curriculum is legendary.

Acknowledgments

I have placed the acknowledgments section here, rather than at the beginning of the book, because it flows naturally from the story-sharing section to thank all who have helped to make this book possible.

My heartfelt thanks go to all the participants in this study, including William Doll's colleagues, William Pinar, Petra Munro Hendry, Wendy Kholi, and David Kirshner. In particular, I would like to thank Professor Kirshner for volunteering to contribute a chapter and for his generosity and patience in waiting for me to finish this book. Doll's former students, including Al Alcazar, Anastasia Chebakova, Mei Wu Hoyt, Tayari Kwa Salaam, Stephen Triche, Shreyl Waltman, and Jie Yu, offered remarkable insights into Doll's life, thought, and teaching. Thank you all for permitting me to use your thoughtful words in this book. Special thanks to Donna Trueit, who has helped me extensively, including generously and kindly hosting me in the two summers when I visited Doll. She has made all sorts of insightful connections that I would not otherwise have been able to see.

All the teaching episodes in the book come from the videotaped sessions of Doll's teaching at UBC with Trueit. The following students participated in the conversations that are included in the book: Maya Tracy Borhani, Brenda Davis, Guopeng Fu, Jung-Hung Jung, Patricia Liu, Rasunah Marsden, Pete Train, and Guanglu Zhang. I am grateful to all for their permission to use class conversations and their help in carefully verifying the transcriptions. Special thanks to Jung-Hoon Jung who helped me videotape the class, contact students for permission, and deliver the digital files. He has been such a tremendous support during the process.

My heartfelt thanks go to Mary Aswell Doll, who gracefully responded to my last-minute questions through email. I cannot thank her enough for sharing her unique insights in so timely a fashion. Throughout these years of working on this project, William Pinar has persistently supported me without any reservation. This book highlights William Doll's pedagogy, but both William Doll and William Pinar have been important mentors to me, and the doubling of their different mentorship styles has worked magic for my transformation. In particular, I am grateful that Pinar has taken the time and the care to craft a wonderful introduction for this book. Thanks also go to Molly Quinn, whose work always inspires me, for writing a poetic Afterword.

I am thankful for Christopher Myers's support of this book project and all the staff at Peter Lang who have helped to publish this book. Thanks also go to

Frances Griffin, who has been my professional editor for fifteen years, for all her marvelous editing and timely help, and Jenna Min Shim, whose rigorous scholarly work I always admire, for her final reading and editing of the manuscript.

Since this book has taken me more than seven years to complete, several graduate assistants have helped with different aspects of writing and formatting the book manuscript. Thank you, Jill Martin, Cicely Fanning, and Crystal Themm for all your help to make this book possible! I also thank the College of Education at Oklahoma State University for granting me a course release in the spring semester of 2014 for this book project as a part of the Faculty Scholarship Support Award. Thanks also go to the interim School Head at School of Teaching and Curriculum Leadership, Jennifer Sanders, the interim Dean at the College of Education at OSU, Robert Davis, and the Provost of OSU-Tulsa, Raj Basu, for their timely support with a ten-hour graduate assistantship during the spring semester of 2016. Importantly, I thank my colleagues and friends at OSU, Pam Brown, Kathryn Castle, and Qiuying Wang, among others, who supported me over all these years.

I appreciate my husband, Zuqiang Ke, for his love and extraordinary patience, which have accompanied me through this writing process. My affection and love for my nephew, David Man (Fang Man), and our time together have been such a gift to me throughout his short ten years of life. Because this book is about pedagogy, I dedicate it to David as his wondrous becoming unfolds.

Most important, let me express my deep gratitude to William Doll. After I first communicated from China with him through letters, his passion for ideas and his openness to a stranger brought me to study with him at LSU. My life has been changed in irreversible ways under his mentorship. What I have been able to accomplish would never have been possible without his pedagogy. What has kept me going for all these years to finish this book is my faith that it does not matter what I produce (of course it still matters), he would welcome whatever it became. Even though it is really difficult to see oneself through another person's eyes, he would see the good in the book and not dwell on the parts that he might not like. His attitude toward life and our pedagogical bond that has developed over twenty years have helped me to undertake this process and emerge from it with a partial sense of fulfillment in my effort to pass on his legacy. Thank you, Bill, for everything.

Naturally I don't see this book as *my* book, even though I am the writer, but as a book of shared stories authored by all participants. Pedagogy is about the next generation. William Doll's stories are among those rare gems that the next generation should cherish. Let's listen to his teaching . . .

NOTE

1. For the remainder of the book, I use the name William Doll. He is also referred to as Bill or Bill Doll by colleagues and friends. I have followed the choices of the interviewees in the quotations from the transcription while using formal reference in the text.

· 1 ·

PEDAGOGY OF PLAY

> Our own [postmodern] vision is a good deal more doubtful, inherently filled with problematics, rooted in dialogue and history, and continually remade as we interact *playfully* with ourselves and the environment of which we are but part. (Doll, 1990/2012, p. 152; emphasis added)

William Doll laughed when I told him that the title of the book would be *From the Parade Child to the King of Chaos*. His laughter is loud and long with a unique rhythm; anybody who walked into Peabody Hall where the College of Education is housed at Louisiana State University (LSU) would know that he was around when he laughed. To my delight, he was delighted with the title. I actually came up with it quickly after he told me stories about his childhood. Here is his story about the parade child:

> There used to be parades in town, a small town near Boston. On the Fourth of July or Memorial Day, there were always parades. The parades were always led by the policemen marching first, and then came other people, except there was a little kid who marched in front of the policemen. That was me. I led every parade. Nobody ever kicked me out. Mother used to say, "Should he really do that?" And Father said, "He's okay." When I was seven, or eight, or even six, who knows [at what age]. Nobody stopped me. Obviously when I was a little older, I realized what was going on, and stopped doing that. (Interview with Doll, 2009)

"The King of Chaos" was a term that students attributed to him. When I first came to LSU as a doctoral student in 1996, I noticed that on Doll's desk was a photo of him smiling, wearing his bowtie and a high hat, with his students' signatures around the photo. They had titled it "The King of Chaos" at the edge of the photo frame. Doll told me that students in the elementary Holmes program (elaborated on later in this chapter) crowned him the King of Chaos at a year-end party. Here "chaos" refers to the chaos and complexity theory that Bill had introduced into curriculum theory and teacher education courses—he was a pioneer in that area.

In his highly influential book, *A Post-Modern Perspective on Curriculum* (1993), Doll devotes a chapter to Ilya Prigogine's theory and its curriculum implications. Very few teacher educators at that time examined what chaos theory would mean for education; it is almost counterintuitive because order is supposed to be the target of the pursuit. However, Doll (1993, 1998b, 2002c, 2004, 2012) advocates a different sense of order, a complex and dynamic order at or near the edge of chaos, and a different sense of control, an emergent control open to flux, change, and creativity. Students' love for him and their appreciation of his nonlinear teaching were crystalized in the endearing term "King of Chaos."

As he reflected on his pedagogical life history, Doll pointed out that his journey from the parade child to the king of chaos indicates the transformation on his part of letting go of control (without giving up leadership) and following a nonlinear flow in chaos and complexity. Certainly the passion for life and the spirit of play in the parade child has stayed with him, although as his journey continues, his play and performativity have become more complex with his emergent sense of pedagogical authority. The jacket photo for this book, taken by Donna Trueit, who accompanied him to China in 2000, conveys this playful spirit clearly. It was taken in Shanghai where they started their academic tour with me as their translator. There was a bronze statue of an old man—a figure of the old wise man in Jungian symbolism (Mayes, 2005)—on the street sitting down at a Chinese chess table. When Doll saw the statue, he took it as an invitation to play. He sat down and happily joined the game. Playing with a bronze man in the imagination captures the spirit of his teaching, first as a schoolteacher and then as a teacher educator and a curriculum scholar.

It seems quite fitting to open this book with his pedagogy of play, an element essential to his life, thought, and teaching. It is also important to teacher education, although often neglected in practice. It is not uncommon for teacher educators to introduce innovative teaching models to students yet teach in a traditional and serious rather than innovative way. Pre-service teachers and in-service teachers seldom have a chance to *play with* new ideas in the classroom. As Ted T. Aoki (1990/2005) points out, work and play are usually separated in education with "the rather simple-minded attitude of either work or play, but never, never, work and play together" (p. 357). However, being playful can bring vitality to curriculum and teaching. Teaching, learning, and play go hand in hand.

There is a sense of flow and spontaneity in play that can achieve organic connections between and among instructors, students, texts/subjects, and context. In such play, students and instructors can be so engaged with the task creatively and collaboratively that they forget about time. Play opens up new possibilities and keeps knowledge alive (Doll, 2005/2012). In the field of education, play is usually perceived as schoolchildren using various materials to learn through hands-on activities in the classroom. However, play is not just about hands-on activities; it is also about being playful with subjects, ideas, and relations (Doll, 2012). In William Doll's teacher education classroom, there is plenty of such play—intellectual, social, and spiritual.

Play in Childhood and Youth

> Play and learning fit together nicely (as most anyone who cares for children knows). (Doll, 2008/2012, p. 15)

William Doll credits his sense of play to his father, who was not restrictive but gave him a lot of freedom to explore: "That sense of freedom and being able to be on my own and leading the parade was all quite important to me and my upbringing and the way that I approach things. So I am quite willing to do things my way" (interview with Doll, 2009). Having the freedom to explore during childhood cultivated in Doll the capacity both to negotiate and to hold his ground. He narrates a story of his play:

> One of the things my father did when I was seven or eight years old, certainly before his heart attack, was driving me to his alma mater, Cornell. It was a phenomenally beautiful campus. It was built between two ravines. They are about eighty to one hundred feet deep and the campus is between them. There was a bridge across. I looked over the side, and I took off my sunglasses, and I let them drop. Mother was just out of her mind: "This is something we bought and you are wearing, you are

wasting it. . . ." Father said, "It is okay, Mother, he was just experimenting." And he would also let me wander around here and there, and he would say: "He'll find his way back. He's okay; he will find his way." (Interview with Doll, 2009)

Doll comments that his mother provided him more guidance but she also followed his father's wish—his father passed away when he was young—to allow a strong sense of flexibility. He thinks that his mother was a playful person herself and passed it on to him, not by direct teaching but by example, as she was a humorous and fun-loving person. He recalls, "She would look up at the clouds and say: 'See that: you see the bear there? You see the old woman's face there?' She saw different figures in the clouds and would roar with laughter" (interview with Doll, 2010). Mary Aswell Doll also fondly remembers William Doll's mother, who had "a child-like sense of being in the world and told funny stories about herself, her own misunderstandings. I laughed so hard I almost fell out of my chair" (personal communication, December 30, 2015). Raised by a mother with a laughing spirit and a father who gently allowed and accompanied his exploration, Doll grew up with a playful spirit.

While attending college and graduate school, Doll also enjoyed a certain degree of freedom. He credited Johns Hopkins for allowing him to muse and to follow his own path, and particularly his mentor, Steve Mann, who both challenged him and gave him freedom to explore. Mann allowed him to study the history and philosophy of science for a whole year simply because he was interested in that area (Doll, 2009/2012). In the Hopkins tradition of graduate education, students had to be able to study on their own and handle questions from a variety of disciplines. Doll's doctoral advisory committee was composed of five members, three of whom were outside of education. So he has a memory of a free and flexible upbringing and graduate education.

As I discuss in more depth in Chapter 6, play in Doll's teenage years happened outside of school. He learned many lessons from his work experiences—he started to work fairly young—as schooling tended to be constraining for him and he did not fit into its peer culture. While his parents allowed and encouraged him to play, school experiences were marked by seriousness and lack of dynamics. The combination of the two in contradiction, freedom to explore as a young child and as a graduate student, and confinement to rules at school, paved the way for Doll to become a playful educator.

A Teacher's Play

A serious teacher, though, is a playful teacher—one full of play. (Doll, 2005/2012, p. 114)

The Role of Teacher

As William Doll recalls, he was a serious teacher at the beginning of his career, but he quickly learned that play, rather than seriousness, takes students further in their growth. The playful spirit he experienced as a child recursively came back to him and decades of experiences as an educator taught him that playing with ideas, with words, and with symbols is an important way to learn (interview with Doll, 2010).

"Teaching was an accident for me," says Doll. He studied mathematics and philosophy at Cornell University. Upon graduation, he did not know what kind of job he wanted. For a while, he sold jewelry. Not knowing anything about it, he was able to sell the first piece for $2,500. Then he decided to go through the Teachers' Placement Agency in New Hampshire and started to teach in a small private school in 1954, without training. Leaving that school after one semester, he went to another school in Boston to teach mathematics and stayed there for four years.

In the beginning, Doll used rigid teaching methods and made students recite facts or formulas. But two things happened during that time in Boston that led to long-lasting change. First, on a rainy day, the class was playing a math game, "monkey in the well," which is about calculating how the monkey climbs out of the well (Doll, 2009/2012, p. 15). During the game, Doll found out that a boy who stuttered was able to develop a formula to solve the problem while Doll had to draw a picture to help himself solve it. It really humbled him to realize that a teacher can learn from students and that students with their various talents and weaknesses must be respected. Like Ruth Kane (2007), who learned that "students have knowledge and experience to contribute to my pedagogical development" (p. 65), Doll began to adopt a stance of teaching through the pedagogical question: "What have I learned today from you as students?" Respecting students' humanness and their ability to surpass the teacher (and later, the teacher educator) became important in his teaching.

Second, Doll struggled with students' role in their own learning. He recalled:

> John Holt, belonging to a [Jerome] Bruner group, came to visit my class one day. Students put on their best performance. At the end of it, John says: "That is very impressive, but I am not sure if students have learned the knowledge or they just performed well to please you." The role of the teacher became obvious to me. It has been an issue that I have struggled with all along and still have not quite figured it out. Not an authoritarian, and not simply a popular teacher, it is a fine line to walk. In my early days of teaching, I played both roles. (Interview with Doll, 2009)

At that time, Doll had not yet read John Dewey's work, but the process of trying to figure out the role of the teacher had already led him toward the Deweyan notion of "first among equals" (Doll, 2009/2012, p. 17), in which the teacher plays a leadership role yet works *with* students. In such a pedagogical relationship, the teacher's guidance is not pre-imposed but emerges throughout the educational process. Upon reflection, Doll says that both the roles of an authoritarian and a popular teacher came out of his sense of insecurity. Because he did not have any teacher training, he knew nothing about how to teach when he started his teaching career. He began by being authoritative because that was the common assumption of how to be a teacher. Then, wanting to be popular with students, he was afraid they would find out that he did not know what he was doing. Both roles could provide a sense of safety.

While the lack of teaching preparation was a factor, I think cultivating a sense of confidence is an issue for all teachers, especially for new teachers. Through his experiences with students and colleagues in his initial years of teaching, Doll learned a third way, an alternative to either controlling or being *laissez-faire*. In Peter Taubman's (1992) terms, it is "the right distance" between the teacher and students that is most pedagogical. Later, when Doll was a headmaster in Baltimore, he took away teachers' manuals so that they would not rely on them in teaching, but instead had to build relationships with students who were interacting with subjects and the world around them. Although they complained loudly, those teachers became more appreciative of their students. By working with students rather than imposing knowledge on them or pleasing students, Doll brings a playful spirit to teaching.

Play With Different Subjects

The spirit of play can be introduced into all subjects in different ways. For Doll, one way to work around the restriction of the discipline structure is to reserve a time limited just for play. When Doll was a teacher at Park Elementary School, he participated in the "Play on the Friday" program in nearby

Shady Hill School, at which Jerome Bruner spent some time working with young people. The program started at three o'clock after the school day had ended. For those students who wanted to stay, John Holt, Bill Hull, and other scholars who were affiliated with Bruner would come to explore mathematical problems and different ways of looking at things with the students. Guiding students to play with mathematic relationships using Cuisenaire Rods, they asked questions: What is the relation between multiplication and addition? Do the two overlap? Are they separate? Do they have certain differences in their structure? "The spirit of serious play" infused those explorations (Doll, 2009/2012, p. 15).

Later, Doll argues for reserving half a day on Friday at school for students to play because a rigid school system must be cracked open to vitalize teaching and learning. As David Jardine, Sharon Friesen, and Patricia Clifford (2012) point out, even in the midst of myriad difficulties in today's education, there is a space of abundance in classrooms that invites teachers and students to enter into a relationship with complex and contested meanings through living conversations.

When Doll moved on and taught in Denver in the late 1950s and early 1960s, he carried with him a strong respect for students:

> I taught math at the level of the eighth and ninth grade. I gave students a textbook and they worked together. They basically taught each other, as these kids knew more than I did. They knew enough math that when they graduated from the ninth grade and went to public schools at the Denver High School, the head of the math department made a special class for these kids. I was seeing the brilliance of these young ones and they would decide how they would handle the problems. I was basically sitting there to monitor them and entered in here and there, but they self-taught. And they worked hard to make sure everybody in the classroom was able to handle what was going on. At the time, the state of Colorado was interested in me enough that they asked me to go around the state, and yet I was totally un-credentialed. The New Math I learned, I learned from kids. (Interview with Doll, 2009)

With such a respect from the teacher, students had a sense of playfulness in and ownership of their learning.

William Doll also worked collaboratively with other teachers. He and another teacher ran a Great Books program on Saturday mornings in Denver, using a teaching syllabus designed by experts. The program, which lasted a year or so, was broadcast over the radio; a small group of students participated in a conversation with teachers while other students who were interested listened to the radio as the audience.

We did it on the air because one of our students' father owned the radio station. We would read some of the Great Books. The program had a syllabus we would follow with weekly reading assignment and questions. We picked from the list of books—it was a great list—but we wanted to only choose books and did not want to follow the syllabus. On those Saturday mornings another teacher and I would let things emerge. We would throw out things, let conversations begin, and enter into the conversation, so we became part of the conversation. That was John Dewey's "first among equals." . . . I don't think we would have run the program in this way if I did not have the respect for the intelligence of the young ones, a respect for what they could do.

We went down to the radio station, and we had an hour to have a conversation with the audience. Great Books staff were so impressed that they asked: "Would you run this for the state of Colorado?" And we said, "Sure, all we ask is not to follow the syllabus you put out!" The conversation ended. These books were appropriate for middle school kids. One of the books I remember was *The Red Badge of Courage*. It was about the American Civil War, what it meant to be in the war, and the meanings of patriotism. The kids read the books and we read the books. We had eight students who participated. Our intent was to show what youth could do if they were allowed to read and think on their own. And we would ask them provocative questions: What do you mean by courage? We would not look for definite answers, but look for complexity of thinking about that question. (Interview with Doll, 2009)

Doll and his colleague probed into deeper issues as they discussed the literature with the middle school students: What is good literature really about? What is the story about? What is the theme in that story? What are the other themes? Are there other ways to look at this hero? What is it that makes a hero? The program was a great success, even as it deviated from the Great Books pre-set format.

Doll also worked with the physical education teacher to turn exercises into play, not competitive play, but the play of working together. He described fondly the gymnastics class in Baltimore where he was the headmaster of an elementary school in the 1960s:

And then I developed a gymnastics program. I got a young man from Germany. The things he had these kids do, without a lot of equipment, was amazing. One of the activities he had them do was that they would get together in as tight a ball as they could, each kid pressing, pressing, pressing, to become a small flower, a bud of the flower. Totally together, now the flower is going to blossom. And as it blossoms, all of the petals are all going to open. There was a sense of sacredness as the flower is blossoming. That whole sense of working together was wonderful, and indeed sacred.

Then we did a very simple thing. We built a wooden wall outside. And the wall was about ten feet high, and the kids had to climb over it. Now, how they climbed over the wall was up to them. If the class had ten kids, they had to find ways to get all ten kids over the wall. It was fascinating! You can get one or two over simply by being on backs of kids one way or another, and then you've got to find ways to bring up those who are on the bottom. Among other things, you would need to make sure that heavy ones go up first to pull the other ones up. You would need to have the strongest kids up above who are holding on to their feet for others to lean down to pick up the young ones. It was wonderful for muscular coordination, and it was wonderful for co-operation.... We stayed away from competition games. (Interview with Doll, 2009)

Play in the gym class was for building collaborative social relationships in addition to doing physical exercises. The competition orientation is strong in physical education, and a lot of times the purpose of a game is to win. But Doll and his teacher changed win-or-lose games into experiences of working together. These imaginative and creative games did not require a lot of materials or money, but the children's bodies and minds both got good exercise. Laughingly, Doll told another story:

At the Valley school, we did all kinds of things for physical exercise, but not for competition. One of the things that they did was to stand on their heads. When the state inspector came in, the third grade was primed so that all students stood up to greet him, and then stood on their heads. What a surprise to the inspector! (Interview with Doll, 2010)

I don't know how the inspector responded to such an unusual greeting, but I did have a good laugh with Doll when he recounted the experience!

A Teacher Educator's Play

"Playing with," that which I see as the most powerful for integrating training and initiative, is where one is aware of the rules (of a game or subject) and purposefully pushes against them, maybe to test them or to extend them or to transcend/transform them. In any event, in "playing with" rules and structures, one acquires an understanding of these at a deeper level than merely playing at or in. (Doll, 2005/2012, p. 116)

Play With the Limits

Doll does not let any system or structure or requirement confine his creativity. As Michel Foucault points out, "playing with structure—transforming and transfiguring its limits" instead of "playing inside the structure" (quoted in Miller, 1993, p. 353) requires both recognizing and transcending the limits.

Playing with the limits—not within the limits—opens up new possibilities. Doll takes time to patiently negotiate with the system until its limit gives way to flexibility and improvisation. His negotiation stance with his administrators is: "You give me freedom, and you will find some remarkable work being done" (interview with Doll, 2010). Since he was successful in producing remarkable work, he was able to keep negotiating for more space to play.

Doll did not follow the rule of writing a syllabus at LSU. Because most people—from faculty to staff to administrators—loved him as a person, he could get away with not doing so. Sometimes he pointed to a fractal image[1] and laughed: "That is my syllabus!" And the class would have an interesting discussion about fractals and curriculum. Other times he brought to the class a cup given to him by a student with the words of his four R's—Richness, Recursion, Relation, and Rigor—and an image of chaos on it as his syllabus to discuss the relationships between chaos theory and curriculum.

He was also fond of showing the video *Mandelbrot and Julia Sets* produced by Art Matrix (1990), which demonstrates fractal images in their endlessly moving forms. Doll would ask: "What do you see in the video?" As the images of the fractal on the screen were moving, transforming, and recursively generating similar patterns (yet not exactly the same), students came up with all sorts of imaginative connections. Interestingly, the subtitle of that video is *For Lovers*. Who would make the connection between mathematics and love? It certainly is not a problem for Doll, whose pedagogy of passion (see Chapter 5) was rooted in his play with mathematics.

Doll told me his favorite story about the syllabus:

> When I first started to teach at Louisiana State University in 1988, in the first doctoral seminar, I was quite serious in handing out the syllabus to the students and reading it line by line. At the end of it, I felt there was something going on, so I asked, "What is it?" John St. Julien said: "There is no party in the syllabus!" Later I started a Friday Friends gathering. Students from the Friday morning seminar—and some other faculty and students also joined in when possible—would get together for lunch and talk with one another. (Interview with Doll, 2009)

Sharing food as well as sharing thoughts has become a tradition in Doll's teaching, which has been effective in building a community. As I remember so well, it was also the tradition of the Curriculum Theory Project at LSU: We shared food in classes and had parties every semester. Parties became a site for engaging in "a complicated conversation" (Pinar, Reynold, Slattery, & Taubman, 1995) that was not only intellectually but also emotionally enriching.

Doll is able to set up unusual and eccentric syllabi because being different from others—adopting an "out-of-the-box" approach, as Sheryl Waltman (personal communication, October 27, 2015) phrases it—does not bother him but excites him, knowing he has an opportunity to explore. At the same time, he does not break away from the system, but finds ways of infusing vitality into the otherwise stifling structure. His passion for life allows his improvisational play to be tolerated and accepted—if not encouraged—by others who have worked with him, although once in a while others had to make up his syllabi for accreditation reviews.

Somehow Doll has retained a childlike (not childish!) quality—the inner child in the Jungian term (Jung, 1969)—that appeals to students from different backgrounds and loosens up the boundary of seriousness of adulthood. It draws students out of their ordinary limits to experiment with different things. As Donna Trueit puts it so well, his conversations with her had the effect of "calling me out of myself" to try new things (interview with Trueit, 2010). Playing with the limits of the system, Doll encourages students to play with their own limits.

Play With Ideas

Doll critiques the notion of experience as only hands-on activities in child-centered pedagogy or certain constructivist strategies in the classroom. For Dewey, an experience is not an experience without the second layer of reflection. For Piaget, "intellectual restructuring" is an essential aspect of the child's response to the environment (Doll, 1993). Thus, intellectual activities are key to curriculum transformation, and play is not only play using the hands, but also play using the mind, although the two are intertwined in teaching and learning, and ultimately in a person's existence. Playing with ideas is a distinctive mark of Doll's pedagogy in teacher education. Doll (2005/2012) argues, "To play with ideas, to be precise with one's facts, and to see these facts related to a set of patterns showing us principles is the way, I believe, of keeping knowledge alive" (p. 118). The vitality in knowledge and life cannot be maintained without playing with ideas.

Playing with ideas in the classroom is not only intellectual, but also aesthetic and spiritual as every member participates in "crafting an experience" (Doll, 2004/2012, p. 98). As students are immersed in subject matter and release their own vital energy in directing their learning process, their play with ideas brings forth creativity and deeper understandings. Such play is not only between students and texts or students and the teacher educator, but also among students. The following teaching example[2] is taken from a graduate seminar William Doll and Donna Trueit co-taught at the University of British Columbia (UBC) in the spring semester of 2013. The class was reading Daniel Tröhler's (2011) book *Languages of Education*, which traces the historical influence of the Protestant heritage on education in three countries: Germany, the United States, and Switzerland. A student started the conversation on the meanings of "character":

Patricia: I have a question to throw out on the table. What does "character" mean to you? We talk about it in our discussions, and I find it an interesting word, but I cannot really articulate it well.

Jung-Hoon: Are you referring to moral education?

Patricia: It does not matter; if we talk about the word and use it in the context of education, or *Bildung*, then I am interested in knowing how you think about it and how it can be formed.

Trueit: Good questions! [*Everybody in the class laughs and talks at the same time, wondering about the question. Then there is a brief pause.*]

Maya: When my first daughter was born, we needed to pick a name, but for the first day as I was looking at her, I could not see her with any of the names on the list I had. She was not one of those names. We did manage to name her within seven days, thank God. But things develop and you get to know your child. I guess at a certain point teaching the morals of what is right and wrong comes into the picture. It is an individual thing, although as we talked about that in this class, it is also influenced by a given family and culture, the logic and morals of a culture. Character is different for individuals, and it is formed slowly over time, nurtured in a protective way, but free to be specified.

Patricia: Then what forms a character in your context?

Maya: I realize that I still have not said what character is. Sorry. I think it is environment, what kinds of environment you have: it is a nurturing, loving, open, and accepting environment, or is it like "no, that is not real; there is no fairy!"—if you did that, how would the child's character develop? I sort of blend character, soul, and personality all together. I don't think I answered your question. How about somebody else?

Guanglu: Maybe the purpose of liberal education is to form character, while disciplinary teaching is to train technicians. In fact, in China we had a very long tradition of liberal education before the twentieth century, and its purpose was to form personality and character. Later on, however, Chinese education focused on subject learning.

Pete: Is the idea of character linked to writing? Language forms and shapes us, so writing is a form of expressing that you do so through characters. When there was no printing, you would write, and we would all write by hand, but they would be different due to characters, and

	there would be some trace of individuality in your character. When you write, by moving your body in a certain type of handwriting, the way you are as a character is to be different, as read by people. So when people started handwriting, it used to be fancy, and used to have a trace of presence on the paper. I like to think of character as being something that can be traced of who you are as well as being imprinted.
Patricia:	How interesting . . .
Trueit:	In the definition, character is a mark that is impressed upon you, or upon something, so it has a distinctive, significant mark of any kind, a graphic sign or symbol. That is one definition. And another is a distinctive mark impressed, engraved, or otherwise formed as a brand or a stamp.
Patricia:	Interesting! So the mark is left on the person. But I am thinking more of internally building a character. One has that kind of desire to build character. I think it is slightly different.
Trueit:	I wonder if that is a different sense. I wonder if identity comes into it. If you think about character as something to which you are subject, then perhaps identity formation is in response to that? Maybe there is a tension between identity and character formation.
Pete:	I read something from Merleau-Ponty, and he mentions two ways, imprinted or imprint, so there are two ways of thinking about character formation.
Patricia:	Yes. If we speak . . . I think, in a way, if I only think about the formation of character, it is more about the external imprint, what has been left on one? My next question is: Would it be possible that it is built internally? I think it is more or less the interaction between formation and building.
Trueit:	Yes, interplay. When there is an action, there is a reaction to it [*laughs*]. We are not separate from culture and society, and we are communal beings.
Patricia:	Fu, you wanted to say something earlier?
Fu:	I just read John Dewey's *Events and Characters*, and he explained both concepts. With characters, he commented on people; with events, he commented on happenings. Different from event, character is unique to people, someone different, or someone outstanding . . .
Trueit:	So distinctiveness; there is a mark of distinction . . .
Fu:	Yes, that would be my idea of character.
Doll:	I am fascinated that nobody has picked up the fact that Asian writing is character writing.
Trueit:	Peter talked about writing . . . [*Other students also commented that Peter talked about it.*]
Doll:	Oh, fine. I just slept through that.
Trueit:	Stay awake, would you? [*Laughter from the class*]

Doll: Sorry... there is a distinctiveness in making those characters and the whole art of calligraphy brings forward that sense. Another way is the German notion of developing a person as a unique and distinctive person, so I think it is tied in with *Bildung* and personhood. And that does have that sense of uniqueness. I see character formation as the formation of a unique person and that process of going through it. As you said, I don't think it can be imprinted and I think it becomes superficial if imprinted. We have a whole history of psychology to use that notion of imprinting. But character gets to the spirit of the person, the soul of the person, going back to German romanticism and idealism. How that is accomplished, as Donna and others already said, is through a sense of interaction.

Trueit: But in that part you read from Tröhler, was the notion of *Bildung* not the uniqueness and idiosyncrasy of the individual in relation to the national character?

Doll: Oh, yes, when you get to *Bildung*, and I do worry about *Bildung*, because it is not merely formation of character, but as I read it, it is the character formation of serving the state. That frightens me. I see all the difficulties coming out of Germany during World War Two as aligned with *Bildung*. (Video transcription, April 12, 2013)

From this teaching episode, we can see the flow of playing with ideas between and among students and instructors, and the process of crafting an experience as a class. The notion of character was important for understanding Tröhler's text and Patricia picked it up to open an intriguing conversation. The class as a whole made all sorts of historical, cultural, and international connections through deliberating on this concept. There was also playfulness among students and between the two instructors that brought humor and laughter to the class.

Playing with ideas also means playing with difference. For Doll, one needs to recognize difference, work with it, and play with it. There was an openness among students with one another. Class members made multiple attempts to understand what a Chinese student or scholar was saying and they were patient with one another in clarifying thoughts. A mother's thought on a child's character in formation and abstract thinking of what character means can be quite different but students felt comfortable putting all those ideas on the table for discussion. Such an interplay among different ideas often happens in Doll's classes.

Playing with ideas might be more complicated for pre-service master's level students in the Holmes program at LSU. The public mis/conception of teachers and teacher education is the profession's "practical" orientation—

noted as procedural—and its lack of intellectual substance, and unfortunately such an external impression becomes internalized to a certain degree for teachers-to-be. Not many students in education are expected to read difficult philosophical texts in their coursework, and their encounter with Doll's teaching has not been easy. However, students who experienced the joy of playing with ideas developed a strong sense of confidence. Doll told a story about a Holmes student:

> One of the students went home at Christmas time and met people at a gathering. Somebody asked, "You are in teacher education?" And she said, "Yes." And he said, "That's a totally complete waste of time; it's nothing but gimmicks and playing with kids and blocks and stuff." At the time, the class was reading either Heidegger or Serres so she picked up a book and gave it to him. "Would you just read the first chapter and comment on it?" And he looked at it, "Wow, I had no idea!" And she said, "This is the book we read in teacher education." (Interview with Doll, 2010)

The student's own pride in her own thinking came from her intellectual—and relational—play with the unfamiliar.

Play With Relations

Play is also relational. As Margaret Macintyre Latta (2013) points out, "Play holds 'clues' to human beings' reciprocity with other(s), with the world" (p. 105). In Doll's teaching and program design as a teacher educator, he has always emphasized the role of relationships. From the foregoing teaching episode, we can see that students' play with relations is mediated by their play with ideas. There was a strong sense of community—both intellectual and social—in their explorations. Not only in the context of classroom teaching but also in redesigning an elementary teacher education program at LSU did Doll highlight the role of playing with relations. The process of his playing with relationships for the program to be approved is as intriguing as the program design itself.

In the late 1980s and the beginning of the 1990s when Doll redesigned the elementary program, a major teacher education reform in the United States had been under way, from which the Holmes Group emerged as a leader. The group released its first report in 1986 and quickly influenced teacher education programs nationwide. Although implemented strategies varied, "there was a coherence in the direction of the changes—more study in the academic content areas, more time in student teaching or other internship placements, reduction in the proliferation of education courses" (Fraser, 2007, p. 226). Being criticized as academically weak and professionally irrelevant, teacher education programs searched for new ways of organizing the curriculum and enhancing theory and practice through innovative pedagogy and practicum (Fraser, 2007; Labaree, 2004).

The tension between academic rigor and professional relevance actually existed from the very beginning of the establishment of teacher education as a professional field early in the twentieth century (Labaree, 2008; Levine, 2006). As Labaree (2004, 2008) shows, the emphases on the academic from the university and on the professional from community leaders have been an ongoing battle for more than a century. Arthur Levine (2006) points out, "Since their earliest days, university-based teacher education programs have been the subject of persistent criticism and prejudice" (p. 23). Academic colleagues at the university perceive teacher education programs as having a low academic standard, while school practitioners perceive them as divorced from educational practices. Historically, such a double bind was formed by the two different origins of teacher education programs: universities and normal schools. While the Holmes Group intended to enhance both focuses, elevating both levels is difficult to achieve.

Under public and professional pressure, college of education administrators at LSU were more than willing to reform the teacher education program. With their support, Doll took the name of the Holmes program, but with his playful attitude toward the limits, he adapted the program to his own philosophy of education with the help of several doctoral students (especially John St. Julien). In the 1980s, the push for standards-based and accountability-driven school reform had already begun in full force with the release of the National Commission on Excellence in Education's *A Nation at Risk* in 1984. The purpose of the Holmes Group aligned with this goal of recreating American schools and developing "new, higher standards for teacher education" (Holmes Group, 1986, p. vii). Although Doll was interested in improving the quality of teacher education, he did not pay attention to the demand for standards because the purpose of improving schoolchildren's performance through test scores is at odds with his educational orientation.

Doll did take from the Holmes proposal two important elements: a stronger academic framework and a stronger professional framework, which matched his own philosophy. Unless pre-service teachers understood the subject in depth, he believed, they would not be able to make connections with the patterns in a subject or help children play with a subject. Without a stronger pedagogical preparation, pre-service teachers would not know how to make meaningful connections between students and subjects. To incorporate both frameworks, Doll led the effort to establish a five-year Holmes elementary program (with the addition of one or two summers) that would lead to teacher certification and a master's degree. In this design, sophomores applied to the program, and then in the junior year a cohort group was formed whose members stayed together until they graduated. The fifth year included graduate level coursework and internships at two schools.

For the stronger academic framework, pre-service teachers were expected to take 3000- or 4000-level courses in their concentration area as undergraduate students, be it math, science, or history, rather than 1000- or 2000-level entrance courses. The stronger professional framework was reflected in students' fifth-year experience, devoted to both theoretical studies in education and internships at schools. During the weekday they worked at schools in pairs for four days, working with two clinical faculty members, and on Fridays they returned to the university to engage in weekly reflections on the internship in the morning and to discuss theoretical works in the afternoon.

The relational element of the program was reflected not only in forming a cohort group for these pre-service teachers to create a sense of community but also in putting two students together as partners in the internships. The two interns engaged each other in addition to learning from their mentor teachers. To a certain degree, the politics of such an arrangement no longer put the mentor teacher at the absolute center of authority but gave interns more freedom to explore, inquire, and search for alternative ideas. Furthermore, in order to enhance students' experiences with diversity, they would be placed in two socially and culturally different schools.

Nationwide, one of the most common critiques of teacher education has been the separation between theory at the university and practice at school (Darling-Hammond & Bransford, 2005; Kosnik, 2007; Labaree, 2008). Promoting dynamic interactions between theory and practice in the new program, Doll expected students to reach solid and sophisticated philosophical and theoretical understandings of educational and pedagogical issues through interplay with their internship experiences at school. Such a blending of the theoretical and the practical work would blur the distinction between the academic and the professional frames. Philosophical inquiries related to teaching in schools are both academic and professional and both theoretical and practical at the same time.

However, getting the program design approved by the faculty at the university was a process of playing with tensions and dancing with differences, full of conflicts, turbulence, negotiation, and renegotiation. Redesigning the program necessarily involved removing individual courses, adding graduate courses, and reducing or combining courses in a specific area. Remembering the tensioned time, Doll gives an example:

> So experts come and say: "Where is my course? How can they have a program that does not have my course in it? And by the way, I really need two courses, because you cannot learn anything about my subject with one course; you've got to have two courses!" (Interview with Doll, 2010)

Such territorial fights in Doll's engaged discussions with various stakeholders were all too common during the process. He negotiated his way with faculty through hard work, humor, and relational improvisations. He also had to persuade school administrators who wanted to keep student teachers at school five days a week to help with maintaining the school routine.

Playing with relations is also playing with different ideas in the web of relationships. Without a playful stance in approaching all the hurdles and difficulties in social relations, without an attitude of seeing difference not as an obstacle but as an opportunity, it would have been impossible to get through the system. Playing with relations is also playing with the limits, not to break with the system but to bend it in order to open up new possibilities. Doll's ability to "live with the system while at the same time changing it" (Doll, 2011a, p. 37) is also linked with his willingness to play with perturbation, which I discuss in Chapter 2. Once the Holmes program was implemented, it was a huge success and earned its reputation in school communities because of its graduates' quality work. The elementary education Holmes program still exists today with its fundamental structure intact.

Doll usually taught the fifth-year graduate seminar on Friday afternoons. Reading John Dewey, Michel Serres, or Martin Heidegger in the graduate coursework was undoubtedly challenging to students and it was a struggle for him to help them make connections between those texts and their internship experiences at school. Because of his own teaching experiences in elementary schools, Doll was able to demonstrate the connections through examples and reading texts together with students in a communal inquiry. Remembering her time in the Holmes program ten years earlier, Sheryl Waltman comments that Doll connected "our practice as teachers to our experiences as his students. He taught us by example, modeling the concepts of his postmodern curriculum" (personal communication, October 27, 2015). These Holmes students are in a better position to help their own students because they struggled with complicated texts and they had eye-opening experiences in embracing uncertainty in their own learning.

Doll takes pride in what the Holmes program students were able to achieve. The readings from intellectuals and thinkers can be difficult, but in a year of working together, "their skills in reading and interpretation improve greatly, as does their general intellectual competence. They take pride in their accomplishments and begin to understand life and its experiences from new perspectives" (Doll, 2011a, p. 36). One graduate said that during her first year of teaching, she put Dewey's (1916/1997) book *Democracy and Education* by the bedside to help her navigate the daily grind of teaching (interview with Doll, 2010). Upon graduation, students have developed subject expertise, philosophical depth, and practical pedagogical wisdom. Such rigor is cultivated by playing with ideas, honoring differences in relationships, and promoting interplay between theory and practice.

A strong sense of a community distinguishes the Holmes program at LSU. Professors, clinical professors, students, and teachers at schools all work together (Doll, 2011a). At the end of the fifth year, students make a formal presentation on their research projects. They often present their work in pairs because partnership is built into their studies. As Doll recalls, often three days of seminars were arranged for students' presentations, and parents, teachers, other university faculty, and College of Education administrators were all invited to attend. It became an exciting intellectual, educational, and communal event (interview with Doll, 2010). In addition to presentations, there was also the end-of-year party. The Holmes students crowned William Doll the "King of Chaos" at one of these parties.

The culture of Louisiana, filled with its own paradoxes and traumas of racism (Edgerton, 1991), is also well-known for its hospitality, parties, parades, music, and food—all sorts of enjoyment. Living in Louisiana, Doll enjoyed all the traditions. He loved parties and hosted many himself. It was also in Louisiana that Doll was able to introduce relational play into teacher education in a systematic way. From Doll's reimagining of the teacher education program to his teaching to the Holmes students, his play with the limits, play with ideas, and play with relations are intertwined to release the creative energy of a pedagogy of play in both teaching and leadership.

4 R's, 5 C's, and 3 S's: A Curriculum Scholar's Play

> If we ask where the *new* comes from, the answer lies in interpretative inquiry; inquiry where we and the "other," no matter what or who the "other" may be (person, idea, fact, culture) come into interactive play. (Doll, 2009/2012, p. 22)

William Doll loves to play with words to capture the spirit of ideas. The 4 R's, 5 C's, and 3 S's are good examples of such play. As he explains, the reason for choosing 3, 4, and 5 is to "produce a 3:4:5 right triangle, most Euclidean and modernist, hence reminding me of the arbitrariness of my own work" (Doll, 2009/2012, p. 21). Such humor is a postmodern play with irony. David Kirshner (in this volume) acknowledges William Doll's "postmodern playfulness of shifting frames of reference, narrative structure, and wordplay."

The 4 R's was proposed as an alternative to the Tyler rationale in formulating a curriculum framework. When Doll was writing his seminar book on a postmodern perspective on curriculum, he was expected to propose something practical to demonstrate his philosophical ideas. Playing with "the three R's of 'Readin', 'Ritin', and 'Rithmetic'" (Doll, 1993, p. 174) and the four steps of the Tyler rationale, he came up with 4 R's—richness, recursion, relations, and rigor—to describe a postmodern curriculum framework. It was not meant to be precise but to provide a certain structure for depicting a dynamic and transformative curriculum. "Richness" refers to "a curriculum's depth, to its layers of meaning, to its multiple possibilities or interpretations" (p. 176). "Recursion" refers to nonlinear looping back for generating something new. "Relations" refers to the matrix of pedagogical and cultural relations that support curriculum dynamics both internally and externally. "Rigor" refers to the postmodern search for new interpretations and patterns that are open to indeterminacy and nonlinear movement.

The addition of the fourth R, rigor, was a result of serendipitous play. Doll and his colleagues founded a New York State Foundations of Education Association in the mid-1970s as a local group that held an annual meeting at different places around the state. When he was invited to give a talk and revisited the group in the early 1990s, he told the audience that he had an alternative to the Tyler rationale but it had three R's and he needed to add a fourth R. "What about rigor?" he asked. Here the notion of rigor is not a modern notion of seeking accuracy and hard evidence, but has a French poststructuralist spin; it is about "the purposeful looking for unseen, or hidden, or yet-to-emerge connections, relations, alternatives, combinations" (Doll, 2009/2012, p. 20). The audience's response was so enthusiastic that Doll went back and added the fourth R to the last chapter in his book. It turns out that this chapter has drawn the most attention from educators not only in the United States but also worldwide. A serendipitous play well-received indeed.

Later, Doll (2002c) envisioned curriculum as 5 C's: *currere*, complexity, cosmology, conversation, and community. His playfulness is also combined with intentionality. In the 5 C's, he starts with a sense of the self in *currere* and ends with community, and in the middle is conversation, which provides a bridge (interview with Doll, 2010). Starting with autobiographical experiences, running through a complex structure of curriculum as an open system, connecting with cosmological interdependence and interrelationship, and going through a transformative process of conversing with the text, the other, and the world, the new curriculum paradigm ties together the self and the communal, the human and the ecological, and the dynamic and the structural. At the same time, it is not a seamless whole, as creative tensionality is the driving force for generating new intelligence and new relationships (Aoki, 2005).

Doll also proposed a curriculum frame of 3 S's: science, story, and spirit (Doll, 1998a, 2003/2012). This formulation is more or less simultaneous with the formulation of the 5 C's, rather than clearly sequential: Recursive play is certainly part of Doll's way of generating scholarship. Doll does not perceive his three different formulations as a progression, with one better than the other, but as his efforts to see the world through multiple lenses. Moreover, he takes a stance of self-questioning in which one plays with one's own assumptions and avoids the rigidity of new ideas. After all, the hardness and firmness of modern scientific reason is the target of the Derridian deconstruction (Derrida, 1990/2002).

Combining the logical reasoning of science, the experiential and the relational of story in the personal and cultural context, and the vitality of spirit, Doll also opens their interplay to *mysterium tremendum*, in Molly Quinn's (2001) term. It is a sense of the sacred with awesome, mysterious energy that gives "life or vitality to our being human or metaphorically to a situation or occasion a sense of being" (p. 108). The visual representation of the 3 S's in an open-heart shape that Doll has developed over the years (see Doll, 2003/2012, p. 103) shows the dynamics of incorporating science, story, and spirit into curriculum. While not giving up the analytic strength of science, Doll blends it with the narrative and the spiritual—curving science toward complexity—to breathe in the vital energy.

Spirituality has been important to Doll, although it is not the same as religion in a narrow sense. He has traveled a long way from the traditional, confining religious dogmas he experienced as a child at a Catholic school (see Chapter 2). As he put it himself, he has never been a strong believer in that which we traditionally call God. But he has a fantastic respect for the fact that we are insignificant compared with the cosmos that surrounds us, that dynamic force, the beauty of a sunset, the panoramic sweep of mountains such as Washington's Mount Baker, or spring flowers. There is something around us that is so much greater than we are (interview with Doll, 2009).

Doll's sense of spirituality echoes Parker J. Palmer's (2003) definition: "Spirituality is the eternal human yearning to be connected with something larger than our own egos" (p. 377). His sense of spirituality is ecological and cosmological, influenced by various thinkers such as Gregory Bateson, Thomas Berry, Michel Serres, Dwayne Huebner, and Stuart Kauffman. He cherishes a deeply rooted belief in the spirit of Christ and his teachings as "a guide to how to live a humane, just, compassionate life" (interview with Doll, 2010). Over and over again he emphasizes the importance of humility, and argues that we need to recognize our own finitude, our own flaws, and our own struggles, and we need to be able to move away from the individual self to society and, further, to the cosmos.

> In 2009 when Dr. Doll visited China, he wanted to visit a Catholic church in south China during Thanksgiving time. It took me quite some time to locate a church and I asked the taxi driver to take both of us there. It was a small church and there was nobody there except the priest. There were rows of long stools (seats without the back) in the church and they were quite low. Dr. Doll is a tall person, but he knelt down on the stool and said his prayer. The beauty of Catholic belief, he explains to me afterward, is that the ritual is the same anywhere in the world. Dr. Doll likes to call himself a "heretical Catholic" since he is not of the mainstream. But there is something in the Catholic faith with its ability to connect across differences in time and place and culture that remains important for him. The image of him kneeling down on that small stool and praying has remained one of my most touching memories of his China visit.

In Doll's own words: "Personally I find spirit in John Dewey's sense of *situation*, in Joseph Schwab's *deliberation*, in Stuart Kauffman's *order*, in Gregory Bateson's *difference which makes a difference*, and in Michel Serres's *teaching as an act of humility*" (2009/2012, p. 22; emphasis in the original). It requires another book to fully discuss Doll's theory and practice of spirituality in education, but it is sufficient to say here that this profound sense of the spiritual—or "spiritful" (Doll, 2012, p. 103)—is ever present in Doll's life and work, embodied in his pedagogy of play, pedagogy of perturbation, pedagogy of presence, pedagogy of patterns, pedagogy of passion, pedagogy of peace, and pedagogy of participation.

NOTES

1. A fractal is a mathematical set that displays a self-similar pattern across different scales. It is generated by running nonlinear equations of complex numbers, reiterated countless times through computer programs. French mathematician Benoît Mandelbrot coined the term "fractal" in the 1970s to describe the complexity of the irregular shapes in nature (Capra, 1996). Nowadays, fractals have become more popular mathematical and artistic images. Doll uses a fractal as a metaphor for curriculum to convey its complexity, richness, fluidity, and iterative beauty. See Chapter 4 of this book for more discussion. For more details, see Doll's (2005) co-edited book, *Chaos, Complexity, Curriculum, and Culture: A Conversation*.
2. In almost every chapter, I introduce an episode from the video transcription of Doll and Trueit's team teaching as an example of Doll's pedagogical approach. In the transcription, I use the last names of Doll and Trueit and the first names of students, mainly because students were referred in class conversations by their first names. However, there was one exception. Fu is a Chinese student's last name, not his first name, but I followed class members' references to him as Fu in the conversation to make it easier for readers to recognize the name.

· 2 ·

PEDAGOGY OF PERTURBATION

> By their nature, open systems need challenges, perturbations, disruptions—these are the *sine qua non* of the transformative process. (Doll, 1993, p. 14)

There may not be play without perturbation; perturbation may reach the breaking point without a playful spirit. In the language of chaos and complexity theory, perturbation with the right amount of tension is beneficial for a system's transformation. William Doll's study of Prigogine started in the 1980s when educational scholars paid almost no attention to him. It was accidental, according to Doll; one of his colleagues made a casual comment: "If you like Piaget, why don't you read Prigogine?" So he did, and a new world opened up.

Doll found the process of equilibrium, disequilibrium, and equilibration in Piaget's theory intriguing. In particular the role of disequilibrium in driving developmental change fascinated him: Disturbances in an established equilibrium are "the stimulus or burr that excites organisms to reshape themselves" (Doll, 1993, p. 81). In the context of teaching, perturbations, errors, and confusions that are traditionally dismissed as negative elements become key elements for initiating students' intellectual reorganization to reach a higher level. Reading Prigogine, who argues that transformative change or self-organization occurs "in far-from-equilibrium situations" (Doll, 1993, p. 103), gave Doll another revelation that reinforced the idea that perturbation can be a positive factor in teaching and learning. The role of perturbation in both Piaget's and Prigogine's theories has been influential for Doll's pedagogy, in which perturbation is not only intellectual but also social, emotional, and spiritual. It must be deeply felt.

For Doll, the so-called best practices in teacher education really limit teachers' creativity. In contrast to the frame of "there is one and only one way to do an activity," he believes that there is always another way to look at things. Tom Kieran at the University of Alberta asks students, after they solve a math problem, "Now does anyone in the class see *another way* to do this?" (Quoted in Doll & Trueit, 2010/2012, p. 182). Moving away from the framework in which there is a right or wrong answer, Kieran encourages students to explore further. Influenced by Kieran's questioning, Doll also asks his students a similar question: "Can anybody look at it in another way?" Looking at an issue in another way introduces difference and perturbation, which are not always welcome. In modern Western logic, perturbation is what needs to be eliminated in order to preserve order. But Doll urges educators to look beyond any either/or framework and welcome the roles of disequilibrium, perturbation, and uncertainty.

"Questioning constitutes the piety of thought" is Doll's favorite quote from Heidegger (Doll, 2002d/2012, p. 81). Doll is fond of telling a story about an expert on Piagetian theory. In an intellectual gathering that Doll regularly hosted in Oswego, a guest kept asking questions about Piaget's theory until he finally said, in exasperation, "I am so busy with writing about Piaget that I don't have time to read or study him!" (interview with Doll, 2009). When one cannot tolerate others' questioning, when one is buried in the demand for publication, one loses the fresh eye for new learning, and learning no longer happens even if one can continue to write.

Perturbation in Life

> Wisdom begins with the fear of a unitary solution. (Doll, 2011c/2012, p. 240)

Doll was an active child, unable to sit still. That did not go well in Sunday school at the Catholic church which he started to attend when he was eight years old. He also liked to ask questions, questions that nuns and teachers could not answer and did not want to answer:

> I went to church on Sundays. . . . Right down in the front there was a nun on each row. I squirmed in my seat; I always got a tap from the nun on the shoulder, or was told: "William, sit still; be quiet." And I asked all sorts of questions, and they could not answer them and just wanted to keep me quiet. One of the first questions I asked was: "Is having a war God's way of reducing the population?" (Interview with Doll, 2010)

Doll fondly remembers those pranks and the trouble he and his classmates caused the teachers and principals. In old age, he still has quite a bit of fun in narrating those trouble-making endeavors that gave him temporary relief from the constraining order of formal education. A big problem he caused his military training commander when he was in college is quite telling:

> I was a tall sophomore while most of the others were freshmen, so I was made the one who led the troops down to one end and then turned around and marched back. I did it about three times and then I thought, "This was absolutely silly." So the next time when I marched them down, we did not turn around and we marched into the wall. So groups of students piled on to the wall. The troop was in an uproar and students were laughing like mad. The commander came out of his office from above and he was furious. I took my rifle in my hands and I threw it to him and said: "It is all yours! Goodbye!" I walked the length of the building to the cheers of students. Oh, how furious the marshals were! I was quiet and let them do all the yelling later. I did not say much as I knew that it would eventually end. So I did not get any credit for that training. (Interview with Doll, 2010)

In his refusal to follow the uniformity and non-sense of military drill, Doll caused disturbance to the presumed order. While it did not go well, he also had the wisdom to let the anger of authority be released freely. Although he stirred up things, he did not fully break with the system and finished the training the next year.

Along with these occasional experiences of feeling compelled to disturb the constraining order of the educational institution, Doll was also open to learning new ideas from perturbations that disrupted his own frameworks. One example involved becoming aware of the necessity to go beyond the dualism that is usually deeply rooted in our mind. Doll told this story:

> During my master's program at Boston University, I had a Chinese scholar—I had a number of interesting people there—who was teaching us ways of interpreting the New Testament. I thought that a Chinese teaching the Bible to us was fascinating in itself. But it was the first time I ran into a person who would say one thing on

Tuesday and then another thing that appeared to be quite different on Thursday. We had the class on Tuesdays and Thursdays. I would come from the logical, linear way of thinking I was trained at school to say, "You said this and then you said that, but these two do not necessarily go together." And he would think for a long time and said, "What I said on Tuesday was true, and what I said on Thursday was also true." That moment was the beginning of the dawn of my getting away from the either/or, and I began to think that there are alternatives. [This awareness as] my philosophical background became an important part of all of my teaching in schools from then on. (Interview with Doll, 2009)

This "both/and" framework was certainly a perturbation for Doll at that time, but it was a perturbation that illuminated alternative pathways and inspired him to think differently. Further influenced by Dewey's interactive approach that goes beyond dualism, Doll's intentional search for alternatives rather than one right answer has become a cornerstone of his teaching.

The impact that the social and political turmoil of the 1960s had on Doll is another good example of perturbation that led to his learning and growth. Growing up, Doll trusted in the government, as most Americans did at that time:

When I was at Johns Hopkins, it was Johnson who was the president, and I just assumed that whatever he was doing was good, but slowly those pictures that came back from Vietnam began to shake my faith. Steve Mann, my mentor, very quietly, very gently, suggested that I start to think about what is going on, what it means to be in America, and our relationships with our fundamental values. It took me a while before I became aware that as a citizen of this country, I had to do more than merely follow what our leaders were doing.... It was Steve Mann who helped me to begin to realize that as an American citizen—while I was reading Dewey's *Democracy and Education*—I need to take a more active role. (Interview with Doll, 2009)

Here the role of perturbation brought by social situations and Steve Mann as a mentor is made clear. While Doll had to unlearn some of his fundamental assumptions and to learn how to question as a citizen in a democratic society, perturbation sank in gradually, not as a breaking point. Although Mann himself was a radical thinker, he challenged Doll in a quiet and gentle way, not imposing on his student what he believed to be true. Mann suggested that Doll look at things in a different manner and also allowed him to question the teacher: "You know, Bill, the radicals in my group would benefit from what you have to say, and they will learn from a good, solid, rational person providing alternative points of view" (interview with Doll, 2009).

The teacher needs to reserve a certain sense of doubt about his or her own viewpoint in order to be open to students' fresh ideas. Being open to the role of disequilibrium is needed not only by students but also by the teacher and the teacher educator. This positioning of inspiring but not demanding students to adventure on to less-traveled roads has become Doll's positioning in teaching. When I was visiting Doll during the summer of 2009, Mann was dying and I witnessed Doll's struggles with the sadness of losing him. They had been discussing difficult yet existential questions about life and death, sharing their spiritual beliefs.

However, as Donna Trueit points out, since Mann was a radical thinker and lost his university job (and his academic career) due to his radical views, Doll probably also learned an unspoken lesson, which is not to reach the breaking point (interview with Trueit, 2010). When he did reach a breaking point with authority—he mentioned such an occasion in an interview—he would question whether there was anything else that he could have done to prevent the falling out. While rebellion may appear to create totally new things, whatever one tries to break with usually comes back to the person in another way, because one cannot help but be drawn by the opposite force, as the Taoist *yin–yang* dynamics show.

I prefer Julia Kristeva's (2002) intimate revolt, which connects sensory experiences and new words to give birth to something new, rather than a revolutionary or oedipal revolt that overthrows an old system to build up a totally new one (Wang, 2010). Doll's openness to the role of perturbation in life allows the right amount of tension, difference, and disturbance to play a positive role so that creativity can be released through forming a "dissipative structure" (Prigogine & Stengers, 1984), and intimate revolt can be nurtured to enable recursive transformation.

Perturbation in Teaching

> The teacher must intentionally cause enough chaos to motivate the student to reorganize. Obviously this is a tricky task. Too much chaos will lead to disruption . . . while too little chaos will produce no reorganization. Just the right amount is needed. Because no pre-set formula can tell a teacher what this will be for individual students, teaching becomes an art. (Doll, 1986/2012, p. 143)

Piaget's work has been influential in the field of education. As Linda R. Kroll (2007) recalls, much of the initial effort to use the Piagetian theory in children's school subject learning "focused on what children could not do because they were not developmentally ready. Piaget's theory was characterized primarily as a stage theory that identified when children would be ready to learn certain basic mathematical and scientific ideas" (p. 96). Different from the popular attention to development stages, Doll (1993) insists that Piaget's biological perspective is more fundamental, that it "focuses on the interaction between the organism and its environment, particularly on the way the organism both actively seeks to respond to the environment and at the same time resists any pressures to change its own patterns" (p. 81). Pedagogically speaking, it is the process of equilibration initiated by perturbation or disequilibrium in the classroom that is key to children's intellectual reorganization. "Moments of discovery," in Kroll's (2007, p. 97) words, would not happen without a certain disturbance to the established cognitive scheme.

When he was at the State University of New York at Oswego, influenced by Piaget's work, Doll started to tell teacher groups that their job was to disturb students. That did not go well, of course. In the field of teacher education, a pedagogy of discomfort has been proposed for social justice education in recent decades (Boler, 1999; Martusewicz, 2001; Wang, 2008), particularly in dealing with the difficult emotions that emerge in discussing controversial diversity issues. However, pedagogy of perturbation has seldom been discussed as a general pedagogical orientation. After decades of teaching both in schools and in teacher education, Doll (2012) has developed the connection between Piaget's work and Prigogine's work, in which perturbation plays a triggering role in intellectual development and personal growth.

Brent Davis and Dennis Samara (2002) point out that there are two definitions of structure: architectural and biological. The more common architectural meaning of structure is usually associated with foundations, building blocks, and deliberate planning; reflected in schooling, it implies "order, rigidity, foresight, permanence, and linear progress" (p. 412). By contrast, a biology-based definition "is a reference to emergence. Structure in such cases is always in process" (p. 412). The biological meaning of structure in Piaget with a sense of emergence and dynamics is what is appealing to Doll.

Prigogine's theory of dissipative structure proposes that when perturbations in a far-from-equilibrium system—biological, chemical, or social—reach a critical turning point, the system can "suddenly" reach a new state. The famous phrase "order out of chaos" refers to this process of self-organization (Prigogine & Stengers, 1984). Self-organization refers to the spontaneous emergence of a new pattern at a higher level of development in a living system as a result of the interactions among local components (Doll, 2012; Kauffman, 1995; Prigogine, 1983). In Chapter 4, I discuss further the emergent patterns in Doll's nonlinear teaching, but here we need to question why so much effort is devoted to avoiding or eliminating unpredictability and instability in teaching if perturbation contributes to transformative change.

> *In the autumn of 1994 when I was working at Shanghai Normal University, I read Doll's article on a postmodern curriculum in* Journal of Curriculum Studies. *I was really intrigued by his ideas, although I could not envision how to find a balance in curriculum that would allow emergence yet still provide structure. Perplexed, I wrote to him—at that time I did not have any easy access to a computer, so I sent him a hand-written letter. I cannot say that I was actually expecting any response because I was a total stranger to him, but the "perturbation" evoked by reading his article made me so curious that I wanted to get my questions out. However, to my pleasant surprise, he not only responded but also wrote a long letter to elaborate on his points (later I found out that he also used that letter for a class discussion). I do not have my hand-written letter but I have kept his type-written response. His letter quoted my question as:*
>
> > I agree that curriculum is a process of development by cooperative efforts between teachers and students, but the relatively stable part of the curriculum can't be cancelled. How does one deal with the combination of stability and unstability [sic]?
>
> *Doll responded, "There is, of course, no answer." But he went on to discuss two approaches to the issue: the first is about the essential tension between the stable and the unstable, and the second is about the need for moving beyond the dichotomy between the stable and the unstable. He even took two days to finish his letter as he ended the first part with the note that it was time for him to walk his dog for the night and started the second part as a note the next morning.*
>
> *When I received his letter during the spring of 1995—international letters took several weeks to be received—honestly, I hadn't the slightest idea what he was talking about! The philosophical language spinning out of his "chaotic" mind was beyond my reach. But I was intrigued enough—and moved by his generous response—that I applied to the curriculum theory program at LSU and asked him to be my doctoral adviser. I came to LSU in August 1996. Today I am still grateful to the serendipitous path that brought me to study with William Doll (and other professors) at LSU. I might not have studied at LSU without the right amount of "perturbation" that Doll successfully planted in our initial communication.*

The "Right" Amount of Tension

According to Petra Munro Hendry, who team taught several classes with Doll, Doll loved to throw the class into chaos and he was good at throwing out "the most outrageous questions or bizarre questions just in order to shake people up and get them thinking" (interview with Hendry, 2010). Sometimes he starts a class by a simple quotation from a text, and students respond to the quote from various angles, connecting it with their own lived experience. In such an open discussion framework, any singular explanation, including the instructor's, can be questioned. He does not mind the messiness of discussions, and in fact, the more messy and disorderly they are, the more excited he becomes. Using seemingly simple pedagogical strategies, Doll leads rich and complicated class conversations.

Doll does not set any predetermined content or specific objectives ahead of time and he is always open to unexpected twists and turns during the semester and makes modifications and adjustments. But he does have a general sense of major ideas and frameworks in a class, while letting the specific details emerge with students' participation. He organizes his classes in intentional ways:

> I don't have a specific goal in mind but I have a broad frame in any course—I want us to explore the main frame through all kinds of different perspectives according to people's own interests and I want them to come together in some sort of semiotic tensions. I am a strong believer in the value of tension—not too much, not too little—but you need to have a sense of tension. (Interview with Doll, 2010)

The tension needs to be the appropriate amount—"not too much, not too little"—to make it generative. When perturbation exceeds students' limits there is no positive pedagogical effect. Doll learned it the hard way in his early career as a teacher educator at SUNY Oswego:

> The instructor for a curriculum development course was sick and I substituted for him. There were 250 master's students in the class; the textbook was terrible. I never used the textbook, but made students read Piaget, Bruner, Illich, etc., all those who were stirring up educational thought at the time. I had them work in small groups and make small group presentations. I listened to two presentations and blew up; my phrase now would be: You did not engage the text but gave high school reports! The class was less than happy. On my evaluation, I remember to this day and will remember until I die, one of students made a short comment: What I have learned was how not to teach. (Interview with Doll, 2010)

Over the years since that moment when perturbation was ill-received by students, Doll has developed his art of teaching that stirs up students' new ways of thinking but does not shake them to a breaking point. To use his own metaphor, mixing two opposite elements through gentle stirring but not aggressive shaking is essential for making a perfect martini, and by extension, for making a pedagogy of perturbation conducive to students' growth. Maintaining a dynamic tension "between challenge and comfort" (Doll, 1989/2012, p. 203), he has learned to guide students to take steps in venturing to the uncertain yet still feeling connected. He and Donna Trueit co-taught a class in Canada in 2010, in which a student from Russia, Anastasia Chebakova, was so moved by their teaching that she made the "world's greatest teacher" medal for both of them and presented them with her little note: "You do know how to teach!"

To bring perturbation into the classroom not only does the degree of tension need to be appropriate, the tension also needs the support of other factors in order to achieve a transformative pedagogical effect, such as time, situational pedagogical relationships, and an emergent design.

The Role of Time

Doll (1990/2012) points out, "Time, especially the quality of time, is a key factor. Students must have an opportunity to reflect, to try alternatives, and to disagree" (p. 142). He gives students time to play with ideas and seek multiple perspectives, and he is extraordinarily patient in pedagogically waiting for students to reach their own turning points without much pressure. His teaching can be perceived as a bit slow (interview with Trueit, 2010) but he gives students a dwelling place in the class to interact with others, with texts, and with themselves. With a slower pace, a community can be built, connections can be made, and ideas can be mixed in multiple ways. Doll (1989/2012) suggests that teachers need to combine "flexible time with directed time in the right proportion to allow and encourage [self-organization] to occur" (p. 202). To approach time in a qualitative way, students are allowed to work things out at their own pace to reach new insights.

Tayari Kwa Salaam, a former student, comments,

> Dr. Doll uses the word *perturbation*. That word accurately captures the stage in the growth process when one nestles in the unknown until one "knows." I use the word *percolation* with perturbation because what one does is filter what is not part of the issue, steep what is part of the issue, and distill what settles the perturbation. (Personal communication with Salaam, August 4, 2015)

The word "percolation" captures well the process of dwelling in the issue provoked by perturbation in order to distill new knowledge. It takes time to filter, steep, or distill and to make a pedagogy of perturbation work for students.

Doll (1999/2012) also makes a connection between perturbation and a sense of *becoming*:

> Becoming is being that moves beyond being, away from the centered state of equilibrium to the exciting, dynamic, and perilous state of far-from-equilibrium. This state opens each of us up to the potential that exists: within life, ourselves, and the creative spirit which infuses the universe. Such a state also exists within the primordial nature of the school subjects we teach. There is an aliveness to both these subjects and ourselves (as creative creatures) if we are but willing to explore this realm far-from-equilibrium and near the edge of chaos. (p. 211)

Doll is more than willing to explore such a realm through a sustained time of engagement to allow participants, curriculum, and the universe to *become* in a creative spirit. The notion that school subjects are also *becoming* rather than preexisting is a radical idea. Such a view challenges the whole efficiency movement in American education that is based on a predetermined design and certainty (Doll, 2002c). As Doll's historical studies show, the efficiency movement has been associated with the model of scientific management in which time and resources cannot be "wasted" on the road to a predetermined destination.

Furthermore, Doll's pedagogy of perturbation to unfold becoming challenges the standards movement in teacher education in the twentieth century. In a postmodern frame that recognizes the importance of self-organization for students' learning, "the concept of standards being imposed is meaningless. Standards, along with goals and aims, emerge; they are not imposed" (Doll, 2002d/2012, p. 86). As Roy A. Edelfelt and James D. Raths (1999) point out in their brief historical study of standards in teacher education, the shared assumption of 130 years of effort (1869–1999)—despite differences in recommendations—is that "through a standard-setting process, the profession should identify and encourage the better practices and eradicate the weaker ones" (p. 1). They identify that the historical trend of teacher education has become more "product oriented" and outcome oriented with measurable standards.

Doll argues that because there is no pre-set formula that tells a teacher the best condition for each individual student's self-organizing learning, teaching becomes an art that is open to the unpredictable, the immeasurable, and becoming. "Behavioral objectives with their set predeterminations have no place in this art" (Doll, 1990/2012, p. 143). Teachers and students need time not only to shift toward *becoming* but also to change their mind-set to make such a shift possible. Even a middle school girl understands well how shifting one's mind-set from finding the only right answer to allowing becoming needs time. A Holmes program graduate in 2004 from LSU, Sheryl Waltman learned to become comfortable with uncertainty and she wanted to pass it on to her own students. Sheryl Waltman recalls her first year of teaching at a middle school when she and her students talked about "one right answer" and the need for searching for alternatives. An eighth-grade girl said to her:

> Mrs. Waltman, we have spent the last seven years in school being told that there is one right answer, and that as soon as you found it, the teacher was happy and you did not have to worry about being wrong or thinking *anymore*. Now you tell us that there is not always one right answer and that you want us to keep thinking. You cannot expect us to change overnight! Give us time. (Personal communication with Waltman, October 27, 2015)

This girl eloquently points out that systematic conditioning cannot be unlearned quickly. At the same time, however, by recognizing such a condition, she has approached a critical turning point in her learning. Giving her time, perturbation that has initiated her self-organization process under a good teacher's pedagogical guidance will lead her to achieve a new level of growth. Some students still stay in touch with Sheryl after a decade. A wonderful year of thinking, learning, and questioning has left a long-term effect on these students' lives, just as Doll's teaching has had a rippling effect on his students' lives and their work in education.

Situational Pedagogical Relationships

When relationships between teachers and students are troubled and distrustful, perturbation may only lead to resistance or avoidance. Doll's tremendous respect and love for Steve Mann was the foundation for Mann's challenge, along with the disturbance of that historical time, to sink into Doll's inner landscape to shift his views. Since then, questioning institutions, authorities, and established thought have remained in his philosophy and pedagogical practice. Similarly, students' respect and love for Doll provide the condition for making his pedagogical challenges educative. According to Sheryl Waltman, "Dr. Doll never assumed a stance of power with his Holmes students" (personal communication, Oct 27, 2015) but stayed in the middle of the stream to steer class conversations and allow students to make meanings.

For Doll, nurturing a trusting relationship between the teacher educator and students through pedagogical care and love must be coupled with replacing the centralized control with dissipative control that allows a certain amount of disturbance. Doll (2000/2012) also calls it "self-organizing control" (p. 224). Such a new sense of control can bring discomfort to both teachers and students as the dissipative structure in a far-from-equilibrium system unsettles any preexisting order. As "the first among equals," the role of the teacher is not to impose new ideas but to cultivate students' own sense of responsibility to deal with newness in thought, society, and culture. In this sense, pedagogical relationships that support a pedagogy of perturbation must be both loving and challenging.

Furthermore, the theory of self-organization requires a systemic viewpoint that situates pedagogical relationships in context. To guide a situation under perturbation toward the transformation of the system without going over the edge requires a sense of feeling for the situation. Situations become their own guides, with the teacher playing an important but non-controlling role. In this way, the teacher's control is dissipated into the group, community, and network. Interaction rather than any individual component is the key to a self-organizing system, so a centered viewpoint, whether it is teacher-centered, student-centered, or subject-centered, cannot support dissipative control and its creativity.

Thus, Doll is skeptical about constructivism. In the pedagogy of teacher education, constructivism as a positive response to the earlier behaviorism has become the mainstream during the past several decades. According to Davis and Sumara (2002), there are various types of constructivism in the field of education, but many approaches adopt the variety of a child-centered approach even though it might be a misinterpretation of constructivism itself. For Doll (2008/2012), transformation is a result of interaction through "*situational* self-organization" (p. 122; italics in original) that supports the creative emergence of new ideas and procedures.

Emergent Design

Emergent design is part of Doll's pedagogy of perturbation. Whether perturbation comes "from an external source or from an internal awareness of gaps" (Reynolds, 2005, p. 266), Doll makes the necessary changes to welcome its interruption. For a number of years, he has not written a formal syllabus; instead, he sends notes to students before the class starts. These "love notes," as he called them, serve as guidelines, although sometimes also as perturbation. He has no problem changing the syllabus when he improvises situational responses. As Jean-François Maheux and Caroline Lajoie (2010) argue, improvisation in teaching and teacher education should be encouraged in order to help pre-service teachers prepare for the unexpected. In improvisation, the teacher does not follow the step-by-step procedure but responds to the immediate situation, sensitive to the ongoing lived experience in the classroom.

At UBC in 2013, Doll and Trueit used *Languages of Education: Protestant Legacies, National Identities, and Global Aspirations* (Tröhler, 2011) to help the class understand Western curriculum history in a global context. An objective of the course was to help students develop their own curriculum thought and writing, so each student brought a writing project to the class. The syllabus outlined a sequence of students taking turns leading class discussions on the book and presenting their writings. It was a weekend class, meeting for six hours on one weekend each month.

Doll acknowledged in the middle of the class that the course had shifted from the syllabus that described a particular structure of reading, leading discussions, and writing. The class evolved into its own rhythm, not bound by the pre-set structure. Doll, in his usual enthusiastic tone, said: "This is a wonderful way to teach and learn by allowing such changes!" A few weeks later, feeling that students' writings needed more attention, he suggested another change so that the class could devote more time to students' writing projects. Students welcomed the change because they could explore their own writings in more depth. At the end of the class Doll and Trueit decided to use emails and phone calls for convenience in communicating their feedback on students' writings.

These many shifts and turns did not seem to cause difficulty for the students. A few of them had already taken classes with Doll and were familiar with his style of teaching, so his improvised adjustments did not come as a surprise. In discussions, students appeared quite comfortable suggesting changes, asking questions, initiating topics, or respectfully disagreeing with the instructors. There was a sense of a community of learners who were engaging in open-ended inquiry and participating in the emergence of curriculum and pedagogy. From the following teaching episode, we can see that students were not afraid to bring perturbation to the class:

Doll: Fascinating. Let's go back to you, Brenda. More thought?
Brenda: I am still troubled. I like the idea of German inward focus and American outward focus, and I can see that in the personalities of the two countries. Yet I think both cultures are troubled.
Doll: Oh, yes, yes.
Brenda: That's where I am almost confused by Tröhler because he presents really interesting ideas but he does not take next steps. I don't know if he is expecting readers to do that....
Doll: That depends on what next step you are referring to [*laughing softly*].
Brenda: I am just thinking about the next steps in terms of thought. You have to think about.... Most of us are educating and we would like to think that we are educating in the present moment, but education has that quintessential question, which is the meaning of education. If Germans teach in a certain way and end up with Nazism and defining the *Volk* or God smaller and smaller and smaller, as exclusive: It does not include Jews, it does not include gypsies, and it does not include the mentally ill, and it does not include....
Doll: Yes, all the non-Aryan people! Here you go!
Brenda: But he never says that. Is he expecting us as readers to make that leap by ourselves?
Doll: As I said, he does mention it on page 165, but yeah, you make a good point.
Brenda: Yes, a little bit, but he mentioned it only in passing.
Doll: Yes, it is so true.
Brenda: The American version of it is interesting too. Now the U.S. sees itself as the savior of the world, and it has started to cause serious consequences. It causes a lot of problems for the world ... he does not go into those things. By the end of the book, I don't see.... Well, if I buy into the Protestant explanation, we should be really ditching these ideas or seriously curtail them [*chuckles*]. But he does not really deal with that.
Doll: For me, it provides a fundamental understanding of how we in education are rooted in Protestant values in the U.S., Germany, and in Switzerland. He brings the depth to that which I did not have. The trouble that you have raised, which I agree with, I now find that I can understand better. So the connections you just made between the social gospel and the U.S. as the savior of the world as long as everybody who is Islamic becomes like us are important ones.
Brenda: That is like the reversal of the German version: Yes, we don't want to exclude anybody and we can include everybody as long as they are like us.
Doll: Both are interested in conquering the world in somewhat different manners. I think we need to take reflection upon ourselves about who we are and what we do, particularly as teachers. (Video transcription, March 22, 2013)

In this episode, we witness that Brenda is comfortable with questioning the text and holding her ground in responding to the instructor's comments. Doll carried on a conversation with her and encouraged her own line of thinking while also asserting his understanding of the text. Without a climate of critical inquiry that welcomes questioning and perturbation, such in-depth probing—with its implicit questioning of the instructor who did not explicitly address the issue—may not have emerged. In an emergent design of teaching, students' critical and creative energy transpire in communal inquiry.

The Limit of Perturbation

In his nonlinear teaching, Doll's spontaneity in letting things emerge and taking the opportunity to come up with new ideas, while often welcomed by his students, can be perceived by some students as unwelcome perturbation, as Teaching Scenario 2.1 shows. When students' creative potential is evoked, they perceive it as a fresh, unconventional way of exploration; when students' rigid academic habits developed through many years are challenged, they may find it unpleasant. There were times when Doll's unorthodox teaching style was a bit too much for a few who unhappily felt disoriented. Therefore, holding the right amount of tension in perturbation is not an easy balance to keep in the classroom because the "right" amount is different for each person.

While Doll's perturbation might spin off the edge for certain students, he does not necessarily approach it as a problem. At the same time, he allows perturbation to come back to him whether from the co-teacher or from students (see Teaching Scenario 2.2). He does not see others' questioning as a challenge to his authority, but approaches it as openings to other views. When students come up with more enlightened interpretations, he is readily receptive to different perspectives and learns from them. Perturbation brings "difference" to a teaching and learning situation. Influenced by Gregory Bateson (1972), who sees difference as a positive contributor to the web of life, and Jacques Derrida's (1992) notion of irreducible difference that marks learning as permanent openness to the alterity of the other, Doll perceives playing with difference as an important part of pedagogy.

Teaching Scenario 2.1

William Doll and Donna Trueit are team teaching a class on research and writing at the University of Victoria in 2009. Today is the second-to-last week of the course.

William Doll: I propose that everyone do a presentation in the last class.
Donna Trueit: Oh, that is not in the syllabus!
William Doll: But everybody wants to know what everybody else is doing. Let's do a presentation, and it does not need to be long. Just ten minutes for each student would be fine.

Students just look at each other. Two students in particular have a fit—they have already had trouble with the fluidity and emergence of this class, and they have no idea what to do for the presentation.

(Donna's reflection: It creates what Bill would call "perturbation," while the rest of us would call it a pain.)

Teaching Scenario 2.2

The class is discussing Bateson's book *Naven* at the University of Victoria. William Doll and Donna Trueit are co-teaching it. William Doll is making comments on it.

Donna: But, Bill, that is a dead white European male's perspective . . .
(Everybody in the class laughs. Bill laughs as well, louder than others.)

Student A: How about we read the Naven ceremony from another perspective situated in a female, native society?
Bill: Can you explain more?

(Later in the class, Bill draws on the student's interpretation to inform his readings.)

The two teaching scenarios are reconstructed from my interviews with Donna Trueit. Although not related exactly as spoken, their storylines have stayed true to the stories she narrated. I juxtapose these two scenarios to show the complex effects of perturbation in the classroom. In the first scenario, some students and Doll's co-teacher felt frustrated by his spontaneous proposal, but Doll saw it as a great way to end a class, even though not in the plan. I suspect even today he would still hold the same perspective because making such a change comes naturally to him. In the second scenario, the co-teacher questioned Doll's perspective in a humorous way and a student followed up with an alternative interpretation. Doll welcomed a different reading of the text and later drew upon the student's reading in further discussions. He accepts perturbation to his own way of thinking from the class in a good spirit, just as he expects others to accept perturbation and difference as essential for their learning.

While she is an adherent of Doll's philosophy of perturbation, Donna Trueit does not appreciate too much perturbation being introduced in the classroom, either as surprises directly affecting her as a co-teacher or as too much spontaneous change imposed on students. She recognizes, however, that his capacity "to be in the middle of the flow [and follow up with others] is immense" (interview with Trueit, 2010). Doll has a unique ability to improvise his responses in following and directing class conversations toward a new level of understanding. There is a limit regarding the positive effect of perturbation as various students have different level of tolerances. As Trueit comments, "Team teaching with Bill is a complicated affair" (interview with Trueit, 2010)—team teaching is discussed in later chapters—and her negotiation with his fondness of "chaos" in order to find the right amount of perturbation for students is an unsettling experience itself.

Pedagogy of Perturbation in Leadership

> The key to such organization—dissipative organization, if I may use such a phrase—is to set up (mathematically, socially) or find (biologically, ecologically, cosmologically) frames demonstrating "just the right amount of disturbance." (Doll, 2000/2012, p. 226)

William Doll's ability to live with tensions has come from not only his life and teaching experiences but also from various leadership experiences at universities or in communities. In Chapter 1, I discussed how Doll negotiated in the midst of perturbation with multiple stakeholders to design the Holmes program at LSU. Another illuminating example is his leadership in redesigning the curriculum structure of general education at SUNY Oswego in the 1980s.

It was complicated and challenging to reexamine and change the curriculum of an entire university, but Doll has learned to enact dissipative control in an organization. As a leader, Doll was facing two important tasks in this redesign: First, there was a need for better connections between the coursework at the community college and the university. Second, the first two years needed to provide a solid foundation for advanced work in the junior and senior years. As he recalls, he worked very hard to build those connections with community colleges, connections which had not substantially existed before, and asked community colleges to offer more rigorous courses. This request did not meet too much resistance because they wanted a better transition for their students. It was the struggle within the university that was full of perturbation:

> I went through fight after fight with departments over the quality of the courses they offered in the freshman and sophomore years, and oh my, were they very unhappy that it was a person from education who did not know their disciplines but came in to tell them what to do! It was a wonderful time! And I would then present these [changes] from the university committee to the full faculty assembly and it was my job to argue before the assembly. You know we would go through, item number so and so, write it on the board, and then say, "Here is the committee's recommendation. How do you feel about it?" And oh, would I get challenged and it was a wonderful verbal back-and-forth!
>
> It went on for a number of years, seven or eight years . . . . I had a wonderful Dean. I would go to the Dean and lay out my problems to him and he would listen and look at me straight in the eye and say, "Bill, this is a wonderful opportunity for you to develop a whole new set of skills." He solved none of my problems. He just said: Go out there and work it out. (Interview with Doll, 2010)

Doll laughs as he recalls those years. What he calls "a wonderful time" was actually full of tensions and struggles. It was an open, far-from-equilibrium situation in which perturbation could lead to transformative change. Living with perturbations in unpredictable and uncertain situations, Doll acquired an exceptional ability to accept the role of discomfort, disagreement, and disturbance in creating a new structure. Although navigating through all hurdles was always difficult, he never gave up his basic principles. For instance, he did not accept the rationale of sacrificing the quality of courses in order to increase student enrollment. I am reminded of his story about his father:

> [Father was put] in charge of the Detroit office. That was where I was born. I was born in Detroit in 1931. During the Depression they said to him: "We would like you to go to Pittsburgh and take charge of the Eastern Seaboard for the company, a big American company." Father said, "I'd rather not move." His boss responded, "You know, you don't really have a choice. We are in the middle of a Depression." Father looked at him and said, "Well, I just made a choice. I will clean out my desk now." So there he had a three-month-old child, with no job and no money. One of his employees, who happened to live in Swampscott, called him up and said: "I am getting ready to set up a company. Would you like to be the president?" (Interview with Doll, 2009)

During the interview, I could sense his pride in his father's ability to say no and make his own choice even when not being given one. Not giving in to pressure and insisting on his own choice also marks Doll's leadership style. As Mary Aswell Doll acknowledges, "Bill's father was reputed to be very much his own person" and "Bill is his own person" following his father's tradition (personal communication, December 30, 2015).

When he was a school board member in Oswego from 1981 to 1983, he was well-known for his integrity. Building personal relationships across differences in a community is always important to Doll, as Chapters 3 and 7 demonstrate, but a critical stance to make a better world together cannot be compromised. At the same time, "making a better world is to negotiate with others in a community" (interview with Doll, 2010). Doll's leadership in combining open negotiation and firm principle was much appreciated by the leader of the school board and he was invited to speak to state and national boards of education. He was also invited to be the president of the school board for the second term but he left Oswego before the term began.

Leadership has been reconceptualized under the influence of the new science of chaos and complexity theory. According to Margaret J. Wheatley (1989), relationships rather than individual components, the right amount of chaos, and participation and cooperation are much needed for creating conditions for self-organizing systems to emerge. Leadership that gets in touch with the creative energy must be open to the role of turbulence and uncertainty, yet also hold the ground for a new structure to emerge. Doll's leadership in curriculum change at the university level and his participation in community leadership have not only enabled him to influence the public but also, in turn, enhanced his capacity to teach in unconventional ways yet still be accepted by the academic culture. Pedagogy is a form of leadership in itself; Doll's pedagogy of perturbation both in the classroom and in the larger community mutually benefit each other.

· 3 ·

PEDAGOGY OF PRESENCE

> Relationality is the "glue" (of connections) which holds any system together. (Doll, 2008/2012, p. 28)

Most of William Doll's students have a deep appreciation for his presence in teaching and mentorship that accompanies them through a learning process often filled with difficulties and challenges. Sharing life, sharing food, and sharing ideas with students both inside and outside of the classroom, Doll teaches what he is. As both Al Alcazar and Stephen Triche—both his former advisees—put it, one cannot separate Doll as a person from the class he teaches. Triche perceives Doll as being one with the class: "He does not merely teach us the content; he really teaches himself. He fully believes in the things that he teaches: It is part of him and he is part of it" (interview with Triche, 2012). Such a unity between the teacher and teaching is remarkable, just as Ted T. Aoki (1992/2005) comments, "Good teachers are more than they do; they are the teaching" (p. 196).

Alcazar used to drive with Doll—they lived near each other in the New Orleans area—to Friday seminars that Doll taught at LSU in Baton Rouge. Alcazar enjoyed all those conversations during the drive. He commented, "He is the guy who drives, he is the guy who teaches, he is the guy who invites you to lunch, and he laughs the same way wherever he is" (interview with Alcazar, 2012). William Doll's presence as a person who is genuinely interested in students as persons created a pedagogical bonding that remained unique and powerful for many of his students.

Interestingly, however, Doll admits that as a student he daydreamed a lot, which means that he was not fully present in his studies. He learned much more through his working experiences as a schoolboy, as I discuss in Chapter 6. He does remember each elementary school teacher's name but not what was going on in any particular classroom. He actually received awards in both history and writing in junior high school and high school, but he had no idea how it happened. Not particularly popular at school, he was nominated by his classmates as the only high school student to go to a summer gathering to meet state legislators, so they had great respect for him, although he was totally unaware of it.

Schools were not particularly challenging for him. Upon reflection, he thinks that his teachers did not know the subjects well enough to bring students to a depth of knowledge. He also brought his daydreaming into college when he was studying mathematics and philosophy at Cornell University:

> I was a bit absent in classes there. Once I stood up in the middle of the class. The professor said: "You were late; now you stand up. What is happening?" I said, "I am in the wrong class." The professor asked, "Which class are you supposed to be in?" That I knew: "Calculus." "Well, that IS this class," replied the professor. (Interview with Doll, 2009)

So what happened between being an absent-minded student and a professor of presence? Can the teacher be both absent and present to students yet still achieve good pedagogical effects? Although this chapter's title is "Pedagogy of Presence," it also discusses the dynamics between absence (to allow students to work out problems on their own) and presence (to help students to work through difficulties).

> *My first experience of Dr. Doll's presence was not experienced by me as "presence." When I first came to Louisiana State University as a doctoral student from China in August 1996, he was teaching in Canada. I was, as one could imagine, pretty confused about what to do. My first meeting with him was when I went to the class on pragmatism and postmodernism, a Friday seminar. Before he walked into the class, I could see a tall, thin figure with white hair peeking at the class through the window in the door. Oh, I told myself, that is my major professor. After class, I asked to meet with him to talk about what I needed to do for the graduate assistantship. He said, "Fine, let's walk to the library together and we will talk on*

the way." Somehow Dr. David Kirshner joined us and Dr. Doll started to talk with him while I was walking with both of them. I must have been anxious at that time—How could I not be anxious?—and I became more confused, so I stopped him in the middle of the walk and again asked him my question. He assured me that we would get to that point soon. After we went back to his office at 225 Peabody, he sat down in his chair and said, "Your English is much better than I imagined." I was not much concerned with my English level at the time but I was much more concerned with my assignment as his graduate assistant. I admit that the first month of working with Dr. Doll at LSU was not easy—"perturbation" would be a good word to describe it—as I seldom got any specific guidelines about what I was supposed to do. Then I figured out his ways of doing things, how his presence sometimes could be accompanied by his lack of attention to details, and how this absence could be helpful on many occasions. Obviously the dynamics between presence and absence have worked well for me, and we have formed a close relationship over the years.

Presence and Absence in Life

The experiences of the immature are to develop, grow, be transformed by interaction with peers and adults. This is an all-share process. (Doll, 2002d/2012, p. 97)

When Doll was nine years old, his father had a heart attack and afterward led a very careful life. However, even in such a difficult condition, his father managed to be present with him, and took him and neighborhood children out to play on Sundays:

Every Sunday was my day. And Father and Mother would take me and local kids around.... And Father drove. We climbed over, up, and down the hills, and would go to places with a steam engine. We went to animal farms. We even saw an old elephant and we rode on it.... He had a heart attack when I was nine and stayed in that hospital for months and months and months.... So from my age of nine until he died at my age of twelve, Father led a careful life although he was devoted to business. While he took me out on Sundays, he basically stayed in the car while we went out. He wasn't active at all. The only remedy they had for him at the time was to stay as quiet as you can, and don't let your heart beat too much. (Interview with Doll, 2009)

While Doll did not know his father well, he is acutely aware of his father's presence and influence on him. A sense of freedom to explore encouraged by his father has stayed with him, and he credits both his parents for his courage to do things his own way. Moreover, this presence was not exclusively between a father and a son, but included other children. Doll is comfortable interacting with multiple persons at the same time, which has developed in him a unique ability to build a class community. I discuss the influence of his father's death on him in more depth in Chapter 6, but it is sufficient to note here that his father's presence in accompanying him to the play sites and his father's absence in disciplining him during his teenage years both contributed to what he was as a teacher and later as a teacher educator. A dynamic interplay between presence and absence in pedagogical relationships requires a teacher to provide guidance and leave room for students' own exploration.

William Doll is loyal to others and prefers to maintain his relationships. Mary Aswell Doll recalls William Doll's loyalty and his devotion to his mother when they lived together. She also comments,

> Bill is profoundly committed to keeping alive the connections he has made in his life. Everywhere we traveled he would find friends to re-connect with, even if this meant traveling eleven hours by car on my birthday (!). . . . Bill liked to plan his travel routes to make these personal re-connections, and they were very jolly occasions. Even to this day Bill re-connects with me, refusing to think of "divorce" as an absolute split. (Personal communication, December 30, 2015)

He has kept in touch with his old teaching friends, his early students, and his college buddies who were in his life decades ago. Throughout my interviews with him, I also found that he has maintained connections with many other people, and he has fond memories of everybody. He has moved and lived in geographically and culturally diverse locations, but it was the people in those places and his relationships with them that initiated his moves and sustained his stay in a particular place (interview with Doll, 2010). He has also learned to be present to others who disagree with his own perspectives through listening and conversing with them and building personal relationships with them despite the differences.

While his amazing presence to his friends, colleagues, and students is evident in Mary Doll's story, the irony of traveling by car eleven hours on her birthday also hints at his absence to her. In his later life, the dissolution of his marriage (of more than twenty years) to Mary Doll confronted him with the issue of control. Doll credits his team teaching experiences with women professors such as Petra Munro Hendry, Donna Trueit, and Jayne Fleener for making him more conscious of the inner workings of male control. He has gradually realized his own tendency to "lead the parade" without being asked. "Letting the parade go off" into various directions has been a learning process for him.

Later his marriage to Trueit, who has interacted with his control with gentle assertiveness, has helped him see things in a different way. In this learning process, "Donna has been a tremendous help in asking, actually demanding, me to pay more attention to the other, and to be less controlling" (interview with Doll, 2009). To be present to oneself, one needs the help of others. In this sense, a pedagogy of presence means not only to be present to students but also to be present to the self, and often these two aspects are intertwined. As Jerry Farber (2008) suggests, presence requires being aware of others and what is going on in the classroom as well as the teacher's own energy within the self.

In the classes that I took with Doll during my doctoral studies and in his mentorship after my graduation, he encouraged me to explore freely; control was not an issue. A person interacts with others in different ways. While the conventional belief is that one is more loving to those who are closest, on many occasions we also reveal a part of ourselves to those who are close that we don't reveal to others. The classroom seems to be a special place to Doll, who has played the role of the wise old man in the Jungian sense, guiding his students with a gentle hand.

In Jungian symbolism, the wise old man is an elder figure with wisdom, insight, and judgment who guides a person's adventures in life (Jung, 1969). In the setting of the classroom, the wise old man "guides the student-hero down the road of academic adventure to the destination of profound transformation" (Mayes, 2010, p. 73). In the process, students experience trials and difficulties and, through overcoming obstacles with the help of sage advice, develop confidence and the ability to cross into new realms of knowledge and personal growth. In such a relationship, the mentor's task is to endow students with a sense of their own capacity. When I met Dr. Doll at LSU, he was already in his sixties, and I looked upon him as a wise guide for my intellectual, emotional, and spiritual transformation. He is still such a guide for me.

Presence in Teaching and Team Teaching

> Teaching now becomes less a process of efficient transmission and more a journeying with others on a path of learning engagement and personal transformation. (Doll, 2002d/2012, p. 97)

The role of relation in a pedagogy of teacher education is important and complex (Loughran, 2006, 2013). The teaching–learning relationship is influenced by "other relationships such as those between teacher and student, student and student, learner and subject matter, and context and content" (Loughran, 2013, p. 122). A pedagogy of presence situated in the complexity of relationality is not only about teacher–student and student–student relationships, but also about the presence of the teacher and students in what they are teaching or learning in the context of the classroom, community, and society.

As Rosalie M. Romano (2002) points out, "What is too often overlooked in teacher preparation is the quality of relations with students that is necessary for fostering engaged learning and thinking, and that fosters a social construction of knowledge" (p. 155). An instrumental approach to teacher education often focuses on how to teach in order to achieve desirable outcomes, thus "concepts of knowledge and knowing are linked in contemporary imagination to accountability and control" (Stengel, 2004, p. 151) but not connected to the relational. However, "the very idea of knowledge depends on the presence of relation and vice versa" (p. 151).

Acknowledging the relational character of education, Gerta Biesta (2004) further points out that "this relation is not one where there is a direct input from the teacher into the mind of the student" (p. 21). If the route is not simple or direct, the influence of pedagogical relationship must go through students' presence to their learning process. Charles Bingham (2004) argues, "Educators have a responsibility to see to it that students are not only reactive to the authority of teachers; it must be insisted that students know how to be active vis-à-vis the authority of their teachers and the authority of content" (p. 35). They need to learn how to navigate when there are conflicts between these two authorities.

In Doll's pedagogy of presence, relationality lies at the heart of teaching and learning. He provides pedagogical companionship and sustained engagement to contribute to students' own sense-making and meaning-making. His teaching is indirect in the sense that it is the student who struggles with texts both individually and communally. Students are always invited to play an active role and they are free to criticize texts and raise challenging questions. In this way, students are present to the text and to the teacher, although such a presence on the part of the teacher does not have a straightforward or linear impact on students. Several important aspects of William Doll's pedagogy of presence are discussed in the next sections.

Pedagogical Companionship

Doll's pedagogical companionship in various forms—often in nontraditional forms—nurtures a culture of sharing in which students not only feel free to explore ideas but also feel supported in exploring their own pathways. He does not see acquiring knowledge as the end of education; he sees developing better human beings as the goal. In his pedagogy, sharing ideas happens in many different settings: Class members share lunch after the class to keep conversations going; he walks with a student to talk about various issues; he drives with students to conferences to make presentations; he makes lunch appointments with an individual student to listen to his or her concerns. In all these scenarios, he is present personally with students in their adventures, intellectually and socially. His ways of "walking and talking" and "driving and talking" have been deeply appreciated by many students. Even without a physical presence, Doll is still in contact with students through emails, phone calls, and letters to keep conversations going.

> When I was his doctoral student, it was not unusual, when he got excited about certain ideas, for him to pick up the phone and talk with me. He would answer my questions he had not answered fully before, or point out a new direction that he thought would be worthwhile to explore. I had never experienced such spontaneous, egalitarian, and caring devotion from a professor before, and it was both touching and significant for me to feel that a teacher cared so much about my ideas. As an international student who had come to a foreign country with its cultural orientations often opposite to my own, I was anxious about many things. I quickly learned, however, that I was under no pressure to take any particular direction under Dr. Doll's supervision. I was free to explore, and in my own explorations he continually provided guidance and inspiration. He was always there for me and my growth, especially when I went through any particularly difficult time.

Doll always walks the extra miles for students and yet considers it a natural part of his teaching. As a result of his enthusiasm, students have also become more present to one another. They send their ideas and news of their own lives to the class through emails, and sometimes such communications continues after the class ends. In this way, "knowing is response-able relation" (Stengel, 2004, p. 139) in which teacher and students respond both to texts and to one another in order to know the world as educators. Doll's pedagogical companionship has nurtured students to free their imagination, reach new insights, and open up possibilities, but he does not try to control the outcome.

In the context of teaching in the classroom, Doll is very good at affirming students' comments or progress with enthusiasm. Petra Munro Hendry noticed such an affirmative stance and its pedagogical effects in team teaching several graduate seminars with him (interview with Hendry, 2010). Not only his verbal responses but also his body language showed his appreciation of students' ideas, such as gestures of attentiveness, an exclamation of great interest, or leaning in closer to show his eagerness to listen more carefully to students' ideas. The videotapes of his co-teaching with Donna Trueit at UBC in 2013 clearly demonstrate such pedagogical positioning.

Doll's presence in teaching is also shown in his ability to weave students' interactions in a recursive manner. He remembers each student's interest and during the class interactions, he weaves text, additional resources, and other students' comments back to a particular student's interest. He may come back to a student's comment a week or a month later to build further connections and students' interactions are enhanced by such weaving. His pedagogical attentiveness and deep listening in such recursion make students feel their perspectives matter. In a pedagogy of presence, the teacher and students share "the present moment" and "are implicated in it" (Farber, 2008, p. 218), but such a moment with all its immediacy and spontaneity is also layered by the past and history. A recursive time situated in the present invites students to become participants.

There is a parental element in Doll's deep concern for students. William Pinar commented that students had strong relationships with Doll who "has a parental, in a positive sense, relationship with students" (interview with Pinar, 2012). When I was a student my experience with Doll was that he would get excited about certain ideas and make all kinds of suggestions; he did not get attached to any particular idea, but rather offered ideas and then left it up to students to decide what to do with them.

Stephen Triche jokingly called Doll "a paternalistic postmodernist" (interview with Triche, 2012). In this term, the combination of parental devotion to the younger generation and the fluidity to allow freedom in postmodernism is actually a good match for Doll's style of pedagogical engagement. "Engagement without attachment" to a predetermined goal is an important element for establishing organic relationality (Wang, 2014). The element of non attachment makes engagement both fuller and liberating.

Doll also expects students in teacher education to take the responsibility to be present in their teaching:

> I certainly remember one young lady who did not show up for her teaching assignment at school on a certain morning because she had been partying the night before. I went to her dorm and got her out of bed, dressed, and I took her to her assignment site. I was strong in the sense of meeting one's obligations. (Interview with Doll, 2010)

Doll's sense of responsibility is tied to his sense of presence. One must be there to meet the other and respond to the other. In this sense, a pedagogy of presence requires a commitment to a response-able relation on the part of both teacher and students.

Sustained Engagement

Doll not only engages with students in the class but also maintains his presence outside of the classroom and after students' graduation. Alcazar appreciated those conversations with Doll on the drive to Friday seminars. As they explored different ideas in the car, Alcazar used some of them in his own life when they applied to his situation and questioned them when they did not apply. Doll liked the questioning and used it as a springboard for more conversations. They never lacked a sense of humor when discussing serious issues: "He was laughing and I was laughing and then we realized that it was time to get out of the car and get ready for class" (interview with Alcazar, 2012).

When Alcazar lost both his parents in his sixth year of doctoral studies, Doll stopped by his house on several occasions, to have dinner and to allow Alcazar to share his grief. As Farber (2008) points out, students bring to the classroom "all their old anxieties, all their unease about themselves in relation to others, and about others in relation to themselves" (p. 217). A pedagogy of presence needs to create a supportive condition in which sharing joy and sadness are both permitted in order to establish personal connections and promote personal growth.

Having a walk and talk with Doll brought joy to Stephen Triche, and that is something that Doll has done with many students. One thing that Triche mentioned several times in my interview with him was Doll's patience. By staying with students during their periods of difficulty and giving them time and space to work out their issues, Doll's pedagogical patience enables students' exploration and gives them courage to adventure into the unknown. His engagement with students does not end with their graduation. After Triche began to work at Nicholls State University, he invited Doll to visit his classes every year. He remembered those occasions:

> We brought him in, talked about ideas, and asked for his feedback. . . . [Students] were all amazed. Particularly in my early years, I was developing my teaching style. He was so different from anything that ever happened [in class]. . . . At different times when we would engage in front of students, he and I would go back and forth, and that was always a huge thrill for me, because it was professor to professor. . . . We enjoyed going back and forth in friendly and stimulating ways to play off ideas, particularly when we had different perspectives. . . . I would invite other faculty and

administrators to be part of it, and he developed friendships with some of them, which extended our relationships . . . so it was the continuous interaction that meant so much for me . . . and it relieved a lot of stress I had as a new faculty since I did not quite fit in. (Interview with Triche, 2012)

From this narration, we not only see that Triche continues to receive inspiration from Doll's mentorship, now more at the level of a colleague, and Doll's sustained engagement with the next generation of teachers/scholars; we also see that Triche reenacted some of the key elements of Doll's pedagogy of presence.

Doll's sustained engagement extends his presence to a wider network through his students. My own students had a chance to interact with him when he came to Oklahoma State University to give a speech in 2006. At the time the class was struggling with reading his co-edited book *Chaos, Complexity, Curriculum, and Culture* (Doll, Fleener, Trueit, & St. Julien, 2005). His presence made the book come alive, and what students previously found difficult to understand became vantage points in class discussions after he left. They were most impressed by his spirit and his ability to energize the class over difficult ideas.

I have to admit that I could not pin down how the change in class dynamics happened, but clearly his presence brought a spark and inspiration to my students' intellectual conversations. His personal embodiment of what he was writing created a link between complexity theory and pedagogy that had been invisible for my students. Such a condition of "nothing happened but everything happened" is akin to the Taoist generative dynamics in which a sense of flow is achieved through meaningful connections between and among teacher, student, and text without asserting much effort (Wang, 2008, 2014). Doll told me later that he found students' questions and comments engaging.

Such an engagement can have profound personal influence. Mei Wu Hoyt worked with Doll for a few years as his graduate assistant and then moved on to another university to continue her doctoral program. Their bond was personal:

> When all my family members were not able to make it to the United States to attend my wedding, Bill Doll happily accepted my invitation to take the father role to give me away at my wedding. During my wedding, Bill got to know my husband's grandparents, who were farmers in Louisiana for several decades. Especially the grandpa had tremendous knowledge about land, crops, plants, and Louisiana. Bill met them several times at different places before both of them passed away a few years ago. He read them just like reading a history book of Louisiana. I don't think either me or my

husband had ever had a deep understanding of the grandparents like Bill did. Likewise my grandparents in-law remembered him forever and asked me about Bill whenever I visited them. (Personal communication, July 29, 2010)

Doll's genuine interest in others goes beyond his students and extends to their relationships as well. He is committed to sustained relationships: "The moment I enter into the relationship, I want to stay with it, and I learn as much from [students] as they learn from me" (interview with Doll, 2010). Remaining a student of students (Ayers, 2002) is Doll's approach and he takes special pride when he thinks that a student intellectually goes beyond himself.

Presence and Absence in the Midst of Difference

Doll also knows the art of stepping back to enable students' learning. As Leah Fowler (2005) points out, the archetypal difficulty of teaching—letting students learn—cannot be overcome but must be lived with because students' learning is beyond the teacher yet is dependent upon the teacher's capacity to create educative conditions. In class discussions, Doll is fully present yet at the same time gives students room to explore, experiment, fall down, and get up on their own.

Doll allowed students to take risks without knowing whether the outcome would be the intended one. Triche was able to defend his dissertation and earn his PhD even though his extension time had expired years ago. Doll supported Triche not only intellectually—I remember those times when Triche came to Doll's office and talked about his dissertation for hours—but also emotionally and institutionally. With his negotiation skills, Doll managed to persuade the LSU administrators to grant Triche his PhD, a remarkable achievement for both the student and the professor. Both had taken the risk. What if Triche had not been granted his doctoral degree despite his successful defense of the dissertation? The outcome was uncertain when they journeyed beyond the time limit, and very few would venture so far.

The effects of Doll's presence also go beyond the boundary. He engages difference through presence and contributes to students' own growth beyond his own expertise or cultural background. Anastasia Chebakova's story is a good example. A student from Russia with a major in political science, she took Doll's and Trueit's team taught course in writing research at the University of Victoria in 2009. She said:

The way they set up the class, how it was different in structure, how it allowed open-thinking and everything—that was all great, but you know, you probably could find it in other places. But for me the amazing thing was that they come from the field of education and helped me go as far as possible in my own field, and nobody from my own discipline had contributed to my work so much. (Interview with Chebakova, 2010)

A good teacher's generative capacity crosses the borders of culture, identity, and discipline. Chebakova decided to pursue her PhD not because of her professors in political science but because of Doll's and Trueit's teaching. The magic lies in enabling students to advance their learning, not necessarily in delivering any specific knowledge. Doll's art of teaching lies in both his presence in supporting students' own intellectual struggles and his absence in allowing them to engage in their own inquiry without knowing where they are going in advance, in much the way that his father accompanied him to play but respected his free spirit when he was a young boy. Jie Yu also credited Doll for teaching her the art of dwelling on intercultural bridges during her studies at LSU while Mei Wu Hoyt discussed how his patience helped her learning. Being able to dwell in time and place despite difficulty and challenge is also a form of presence.

Interestingly, international students' high praise of Doll's teaching is complicated by African American students' relationship with him. To be open to different cultures in one's own town (Porche-Frilot, 2002; Wang & Olson, 2009) is not the same as being open to a foreign culture such as that of China or Russia. Because of the effects of racism, his gaining African American students' trust might be more difficult. Although "learning from the other" was still Doll's approach, it did not always work for those students. In the interviews he emphasized his learning from African American students at LSU and acknowledged that they challenged his view and his Eurocentrism.

As a privileged white male, Doll had his own blind spots and could be unaware of the significant role of race or gender for his minority students, which made a few of them skeptical, but his pedagogical commitment to these students was strong. Some African American students built relationships with him and invited him into their lives. He considered being invited to LaVada Taylor-Brandon's wedding "a phenomenal honor" that he would "never ever forget" (interview with Doll, 2010). Another African American doctoral student, Tayari Kwa Salaam, came to gatherings at Doll's house to sing songs and deliver poetry, captivating the whole audience. He learned from her that "the African-American idiom, 'I be' instead of 'I am,' has connotations not found in the latter more grammatically correct phrase" (2012, p. xviii) and has always been amazed by her vital spirit as a person. Likewise, Salaam affirms that "Curriculum Theory changed my life for the better" and she remembers one particular story:

> I remember how I was sharing in a class seminar early on in my graduate school experience. When Dr. Doll heard I was regurgitating the author's thoughts and theories, he stopped me and asked, "But what do you think?" I had not realized I wasn't sharing my own thoughts and theories. (Personal communication with Salaam, August 4, 2015)

Doll always encourages students to follow their own line of thinking and their own paths. To claim such an agency helps minority students to be more self-affirmative. Salaam also commented, "Dr. Doll was part of the Curriculum Theory Department that was made up of professors who as a group affirmed our brilliance. As African American doctoral students, this may have been the first time we were told this" (personal communication with Salaam, August 4, 2015). Doll always tries to see the good in every student and truly has faith in their capacity. Doll believes that difference plays a positive role and approaches his minority students' presence as enriching his life.

Team Teaching

In teacher education, it is generally acknowledged that collaborative teaching, including team teaching, is beneficial for students' learning, by modeling for students what collaboration means in the context of education, providing complementary expertise, and encouraging mutual learning between professors (Kluth & Straut, 2003; Nevin, Thousand, & Villa, 2009). Ann I. Nevin, Jacqueline S. Thousand, and Richard A. Villa (2009) argue that "how professors perceive each other and interact with one another is a neglected aspect of university life and should not be ignored" (p. 572). Just as students in K–12 schools and in teacher education are encouraged to engage in cooperative learning, peer interactions among teacher educators also need to be encouraged. In this broad context of collaboration, team teaching is defined as a partnership of educators "who share responsibility for planning, teaching, and assessing the progress of all students in the course" (p. 570). Teacher educators engaging in team teaching fully collaborate in the process of teaching.

Doll has remained enthusiastic about team teaching in teacher education. Although he has not adopted an instrumental approach to planning, implementing, and assessing, his advocacy for team teaching is compatible with his pedagogy of presence and philosophy of participation (see Chapter 7) that invite everybody to be part of the circle. Since it is a circle, it cannot be a hierarchical arrangement in which the teacher educator is the sole authority in the class. The mutuality between teacher educator partners as well as their presence to students encourages students' engagement with learning.

Interestingly it is through team teaching—particularly with Donna Trueit and Petra Munro Hendry—that Doll became more aware of the blind spots in his interactions with others:

> I am not nearly as sensitive to the other as I should be. I hope I get a little better at that. Now and then people get nervous at my blustering enthusiasm. With Donna, I have learned to become a bit more respectful of the other person's point of view. It is not that I did not mean to be respectful, but I just kind of overlooked it. I go on with my own way. (Interview with Doll, 2010)

Trueit recalls one of their best collaborative teaching experiences. The topic was about research and writing in an interdisciplinary class at the University of Victoria. Following emergent curriculum design, "it was set up in the beginning to be very fluid. Students accept the fluidity, particularly because they had their own research agenda and we worked with all of them both individually and as a class" (interview with Trueit, 2010). As a result of their team teaching, nine of eleven students published or were in the process of publishing papers at the end of the class. Trueit comments, "That is the kind of teaching through which we had developed this idea of thinking complexly" (interview with Trueit, 2010). Trueit and Doll also published an article based in part on that class (Doll & Trueit, 2010/2012). While the teaching effect was amazing, and they continued to work with those students after the class ended, for Trueit it was not a smooth process.

For Trueit, part of the complexity of team teaching with Doll comes from his "great difficulty in sticking to any kind of structure" (interview with Trueit, 2010). For instance, Trueit developed the general outline for a new class they created and they had agreed to follow a certain reading order for the class. However, later Doll wanted to change things: "How about this? How about we do that?" Doll's spontaneity clashed with Trueit's preference for organization. Although she favored flexibility, she did not like to change things as quickly as Doll did.

Trueit and Doll come from different backgrounds. Trueit grew up in the country and considers herself a "farm girl" who knows a great deal about practical living, whereas Doll grew up in an (upper) middle-class family with the privilege of enjoying "high" art or culture. Trueit's background is in nursing; as a former nurse who specialized in providing care, she is adept at picking up unspoken clues from others. Trueit is more attuned to different aspects of students' perceptions, particularly the negative side, whereas Doll tends to be enthusiastic about the positive side. She, more than Doll, is able to relate to students when they are in distress. Aware of his own limits, Doll also relies on Trueit to reach students that he does not know how to reach. See Teaching Scenario 3.1.

Teaching Scenario 3.1

After the class, Doll asks Trueit how the class went.

Doll:	Dear, how did it go?
Trueit:	How did you think it went?
Doll:	Wonderful! A great class!
Trueit:	I am a little concerned about those two students who sat in the back of the class.
Doll:	Really? Tell me more about it.
Trueit:	They only spoke to each other but not to the class. I wonder what we could do to make them more comfortable to share their ideas with the class.
Doll:	That young man should be okay. He will catch up. I have no idea what to do with that young woman, though. Will you talk to her?

At the same time, Trueit also sees their difference as complementary in teaching: The depth and breadth of Doll's philosophical understandings provide a cornerstone for intellectual conversations in the class whereas Trueit's attention to details and organization gives students a tangible structure with which to work. Trueit also calls the class's attention to Doll's perspective from a particular positioning with humor, as we see from their interactions in Teaching Scenario 2.2 in Chapter 2, to open alternative readings of the texts. William Pinar was impressed by their "intellectual dynamics and what an intellectual pair they are. Donna knows how to negotiate with Bill in some way. She always knows how to work with tension" (interview with Pinar, 2012). The fact that they are a loving couple who teach together also brings to the class a unique dynamic that sustains a positive energy.

In team teaching with Doll, Hendry pointed out the ways in which each of their positionalities granted them different types of privilege (see Teaching Scenario 3.2). The relationship between Hendry and Doll was not only pedagogical but also personal as they shared many experiences over the course of twenty years at LSU that included both immense joy and grief that contributed to a growing sense of a shared commitment to the spiritual dimensions of life. Hendry was comfortable with, and not intimidated by, co-teaching with him (interview with Hendry, 2010). Such a relationship built a foundation upon which Doll was comfortable with the differences they had and treated them as a good opportunity to discuss things further as a class.

Teaching Scenario 3.2

William Doll and Petra Munro Hendry are team teaching a graduate seminar.
Doll: Why don't we give up on having a syllabus? We really don't know where the class is going!
Petra: You could get away with that because you are a "Full" professor, but not me; I still have to answer to the Chair of the Department who requires that we submit a copy of our syllabi. Also, don't we have an obligation to the students? Shouldn't they know what is expected of them? Whose interests are being served by not having a syllabus?
[Students laughed; so did both instructors.]
Doll: But what is the use of it?
Hendry looks at students: What do you all think?
Student A: I don't know. Probably I would be lost without knowing what to follow.
Student B: How much flexibility can we get from the department policy?
Student C: What would it be like without a syllabus?
Doll: The question is how much we can rock the boat....

The mutual presence of co-instructors—not without disagreement—models for students that there can be different views, but collaboration can benefit from differences shared in an open, conversational way. Hendry thinks it was important for her as a junior female faculty member to collaborate with a senior male faculty member in such a way that a woman had an equal and strong voice in the process. To demonstrate that "it was safe to have differences" (interview with Hendry, 2010) in the classroom between two professors embodied a mode of pedagogy that can address gendered biases in critical yet playful ways.

As Joel Westheimer (2008) points out, learning among colleagues in a teacher community through collaborating in teaching and in their inquiry into pedagogical practices has gained increasing attention in the past twenty years, but teacher education programs have not made substantial changes to support such a community. To incorporate such an element of professional collaboration and co-learning in teacher education, it seems to me that teacher educators not only need to discuss the issue, but also need to embody collaboration in their own teaching. The collaborative yet critical team teaching relationship between Hendry and Doll in the context of communal intellectual inquiry with students on a regular basis demonstrated a pedagogy of presence that supported a teacher learning community.

In team teaching, Doll works with others who are from different intellectual or social backgrounds because he accepts their difference and does not see it as an obstacle to his own preferences. When co-teaching a course on cognition and instruction with David Kirshner at LSU, Doll was clear that they were from different orientations: For Doll it was a biological viewpoint whereas for Kirshner it was about artificial intelligence. According to Margaret R. Letterman and Kimberly B. Dugan (2004), interdisciplinary complementarity in team teaching has great potential for enhancing students' learning because students also come from different intellectual and disciplinary backgrounds. Interactions between different frameworks helped to clarify each position and students had a chance to understand different viewpoints within the context and develop their own approaches.

Doll and Krishner had (and still have) a strong respect for each other (see Krishner's contributed chapter) and out of their respectful interactions, difference became a contributing factor to enrich the class conversations. Doll believed that the presence of each professor to students could be strengthened rather than weakened by team teaching. He particularly appreciated Kirshner's independent and creative spirit along with his ability to play with subject matter, ideas, and the self, and they could get a good laugh from each other's ideas without feeling offended.

The collaboration between Paula Kluth and Diana Straut (2003), two teacher educators, one a special education professor and the other a general curriculum professor, demonstrates a similar dynamic in which they discuss their differences in class when they disagree with each other. In modeling constructive debates, they hope to encourage students "to experience and participate in productive disagreement, critical discussion, reflective practice, and scholarly growth" (p. 237). Students appreciate the dynamics generated by respectful and critical dialogues between two teacher educators.

Students have responded positively to Doll's team teaching. Anastasia Chebakova had to think about it when I asked her about the team teaching aspect of Doll's pedagogy. She paused a while before formulating her answers, partly because Doll and Trueit seemed to work together really well from a student's viewpoint. Finally, she said, "They are different and they sometimes have different views. They formulate ideas differently and see things from different angles. But they modeled engagement for us, and it is nice to see their communication with each other" (interview with Chebakova, 2010). When it comes to individual feedback on writing, Chebakova noticed that Doll preferred to look at the big picture or the whole idea whereas Trueit would pick up details that he did not mention. When she talked with them individually, they paid attention to different things, and upon reflection she realized she would go to Trueit with certain questions and go to Doll with others, although it was not conscious on her part. However, she saw such a difference as a good thing.

Interestingly, in Wendy Kohli's recollection of her work with Doll on the Holmes program at LSU, she also mentioned a division of labor:

> We worked well together for the Holmes program. I am not a confrontational person, and I did some sort of translation work for him to connect with teachers and took care of some details. But he was the designer of the program. It was a gendered thing, I guess, and there was a sort of natural division of the labor. He is a person of warmth. The program was running well. (Interview with Kohli, 2014)

It does come "naturally" to Doll, for example, who assigns students wanting to explore the poetic aspect of education to work with Trueit, because he believes that is her area of expertise. There is a gendered aspect in the division of labor, although he does not necessarily acknowledge that. Doll's presence in team teaching, in this sense, is not without absence, because he does not want to deal with what he perceives as things that he is not good at. On the part of students, however, taking advantage of each professor's different strengths is beneficial because they can explore more possibilities and venture into unconventional areas.

Theory and Practice in Mutual Presence

> The issue of theory vs. practice is the false issue. How these two interact is the practical problem we need to face. This is our challenge as educators. (Doll, 2002d/2012, p. 86)

In Doll's pedagogy of presence, theory and practice are mutually present to each other. He practices what he theorizes and his theory is crystalized from his practice; his theory and practice feed into each other. He prefers the term "curriculum thinker" over "curriculum theorist" because of the implications of the term "theorist" as being separate from "practitioner." He believes that those years of interacting with children in his life were immensely important for his formulation of curriculum and pedagogical theory. He changed his teaching style because of the influence of such interactions:

> In my very first years of teaching back in the early fifties, I used to require students to do some very definite performances: One was to recite the multiplication table, and another one was Latin words. I had no use for a student who stuttered. From my perspective, he just needed to overcome the stuttering and I expected him to do the same thing as everybody else did. And then as I said, at recess one day, we were playing Monkey on the Wheel, I realized he was intuitively brighter than I was. He could do it in a formula—symbolic representation in Bruner's term, while I had to draw myself a picture.... So that first year of teaching from 1953 to 1954, I began to question the way I was operating. And started to look for ways for students to explore. (Interview with Doll, 2010)

Even though Doll had a playful childhood, that sense of the playful was not directly translated into his early years of teaching at school, partly because school had been a constraining place for him when he was a student. Studies show that new teachers tend to begin their career by teaching the way their teachers taught them (Pratt, 2011). However, Doll's experiences with children as a teacher had changed his way of teaching. Within five years of teaching, he had a set of devices for encouraging students to explore, to play, and to grow. Later, his reading of curriculum thinkers refined his teaching approaches and enabled him to draw connections between theory and practice at higher and more complex levels.

Doll points out that the major thinkers he draws upon had extensive interaction with children: Dewey ran a school, Piaget studied kids, and Bruner worked with children. That one is able to do what one is preaching about is important for him. In his highly theoretical book on a postmodern approach to curriculum, he also presents 4 R's as practical landmarks for educators to use in their own practices (Doll, 1993). He theorizes about nonlinear teaching and complex thinking; he also practices it in his own teaching (Doll, 1999/2012; Doll & Trueit, 2010/2012).

As a philosopher, Doll has a passion for intellectual inquiry and brings that passion into the classroom to share with students. At the same time, he is fully grounded in practices of teaching, pedagogy, and curriculum. As a teacher educator, Doll chose to be present in practice by modeling to preservice teachers what could be done at school. During the Oswego years when he taught math methods to undergraduate students, he had his students watch him teach at the campus school on Fridays. When coordinating student visits to schools in the community, he felt that it was important not just to talk about how to teach, but to show it. Nel Noddings observed his teaching:

> She watched and offered an amazingly good insight. She said, "Bill, you are a marvelous teacher. You have your students here in this campus school doing very interesting things. But this is you. It is you and your relation to the student; this is not a method that others can necessarily use." (Interview with Doll, 2010)

Her comments reveal that a good teacher educator embodies what he or she teaches, and students' task is not to learn a specific method, but to find ways to build meaningful relationships in their own practices. Methods can vary depending on the teacher and the students, what situation they are in, and what can work magic in a specific classroom with a specific subject. Doll's personal presence in teaching schoolchildren made the connections between theory and practice come alive to students. When he supervised students' field experiences, he also found ways to teach, and if the teacher let him, he would teach that class for ten or fifteen minutes. He did not just sit in the back of the classroom to take notes but entered into a conversation with the teacher, the student teacher, and students in the class. As he was teaching the children, he simultaneously talked with the teacher and the student teacher. He worked to develop good relationships with schoolteachers and administrators.

The interaction between theory and practice in mutual presence is not intended to serve traditional learning purposes but rather to craft students' pedagogical capacities for teaching differently. In Jeong Suk Pang's reflection on Doll's teaching at LSU, particularly in the math methods course for preservice teachers and a master's level course for schoolteachers, she observed "a close relationship" (in Doll, 1999/2012, p. 219) between Doll's postmodern theory and teaching practices. Her understanding of nonlinear teaching came from his practice of it in the classroom, and Doll's teaching helped her "conceive theory in practice and of practice in theory" (p. 219).

While the critique of teacher education separating theory and practice has been historically prevalent (Korthagen, Loughran, & Russell, 2006; Russell & Loughran, 2007), Doll has adopted an embodied and interactive approach to the relationship between the two not only in his own teaching but also in his design of the elementary Holmes program at LSU. However, he does not share the purpose of teacher education reform. Arthur Levine (2006) argues that teacher education programs

> should educate teachers for a world in which the only measure of success is student achievement. . . . [These programs] are now in the business of preparing teachers for a new world: an outcome-based, accountability-driven system of education in which all children are expected to learn. (p. 104)

Such a vision of teacher education reform aligning with school reform—Pinar (2006) calls it "school deform" (p. 118)—is at odds with Doll's consistent critique of an instrumental approach to education (Doll, 2002c, 2012). What Doll has been able to achieve through the mutual presence of theory and practice in his teaching and teacher education program design demonstrates another pathway, an alternative to the behaviorism-oriented, outdated perspective that has unfortunately reemerged in the recent decades of educational reform. For Doll, students' engagement in learning and capacity for "crafting an experience" (2004/2012, p. 98) are more important than measurable learning achievement.

Doll (2004/2012) refers to the immersion into a particular subject, topic, or situation while being summoned to respond as a whole being as "crafting an experience": "It is this process of interactive doing, undergoing, and responding which turns experience into *an experience*" (p. 99; italics in the original). In other words, a student's presence to learning includes both being absorbed into an experience and acting upon the world—the simultaneous process of taking in and responding to. In helping students to craft an experience in the classroom, teachers are also crafting an experience of teaching for themselves. In crafting an experience, students' engaging and experiencing necessarily involve both theory and practice not as separate but in an embodied relationship. The teacher's presence in teaching calls upon students' presence in learning.

Most Holmes students had not had much experience in reading difficult philosophical works in their previous coursework. As Sheryl Waltman—a Holmes student of 2003–2004—recalls, she knew before the class that Dr. Doll would challenge students (personal communication with Waltman, October 25, 2015). The Holmes students were advised at the very beginning of the program that they would be intellectually challenged in their learning, and Doll had a reputation for doing just that. But Waltman found his "out-of-the-box" approach stimulating. In teaching, he used real-life experiences in a nontraditional way—such as counting from left to right in order to get the dollars right at the grocery store (Doll, 1993)—and these examples helped Sheryl understand the significance of questioning the conventional.

Interestingly, Waltman's favorite book during the class was Michel Serres's (1991/1997) *The Troubadour of Knowledge*. Before she returned to school for her bachelor's and master's degrees in education, she was a "stay-at-home mom, then worked in sales and marketing for sixteen years" (personal communication with Waltman, October 25, 2015). With her nontraditional background, she still formed a relationship with this philosophical and poetic text. She did her own research on the rich language that Serres used and looked up some allusions. With the help of the class discussion, she could "see" the relationship between the text and what was happening to students in the context of school, family, and community.

As a philosopher with the practical wisdom of a practitioner in education, William Doll was able to navigate the difficulty by engaging students' presence with theory through finding points of connections, offering his own examples of playing with children, and weaving important themes in a nonlinear manner. There was no pedagogical demand for students to grasp any particular concepts or theories, and what he did was to encourage students to interact with the text using their own life experiences as well as field experiences at school. As Sheryl Waltman points out:

> Dr. Doll set before me a course of thinking and learning, of possibilities, of the beauty and order in chaos. He never told me what or how to think. He merely paved the way and left me to make the choices—and the path I've chosen has made all the difference. (Personal communication, October 25, 2015)

Philosophical wisdom and insights benefit students in the long term as they become lifelong learners, cultivating a broader view of life and education and enhancing the endurance to deal with difficulty. A Holmes graduate who put Dewey's *Democracy and Education* (1916/1997) by her bedside during her first year of teaching is another good example of the impact of Doll's teaching on students' work.

In class discussions, Doll guided conversations while also allowing them to go in their own directions. But when sparks of connections happened, he seized the opportunity to highlight what was happening. He was also always enthusiastic and students were often persuaded by his passion for their learning. In the back-and-forth movement between philosophy as texts and experiences at school, Holmes students made sense of intellectually rich and challenging materials. Waltman recalled that Dr. Doll "had a talent for bringing the discussion around full circle to connect with the internship experiences we brought up" (personal communication with Waltman, October 27, 2015). Such weaving between theory and practice for new insights and pedagogical action is an indicator of a powerful teacher education pedagogy (Thompson, 2006).

It is important to point out, however, that the mutual presence of theory and practice does not mean that the connection between the two is seamless; it is the negotiation with the difference or even the tension between the two that marks Doll's dynamic approach. As Doll expresses so well about what clicked for the Holmes students, "It is the engagement with the text; it is the engagement with students. It is a sense of being present with them, with [instructors] being immersed in what students are doing" (interview with Doll, 2010). I think this succinctly summarizes his pedagogy of presence.

· 4 ·

PEDAGOGY OF PATTERNS

> When we shift our gaze from the particular to the interconnected and to the generativity of all life and creation, a sense of awe arises. It is this focus, rather than our own personal salvation, I am advocating we adopt—a focus on the "pattern which connects," rather than on the individual objects connected. (Doll, 2002b/2012, p. 36)

Playing with patterns has been a life theme and a pedagogical orientation for William Doll. In childhood, he had been encouraged by his parents, particularly his mother, to see patterns, and this "eye" for patterns has stayed with him in teaching. In his own words:

> When we, as a family, took a Sunday drive, Mother encouraged me to look for patterns, sometimes as we crossed bridges, more often as we looked out the window at clouds and their shapes. The same would be when we hung out wash to dry in the blowing wind. I liked the wind blowing and Mother would look to the sky and see imaginary creatures in the clouds. We lived on a hill, overlooking a bay. The wind blew, the clouds flew, and Mother encouraged my imagination as to the shapes—often animals—the clouds took. I was probably seven or eight or nine years of age. (Personal communication, November 11, 2015)

Doll also recalls that as a teenager he walked a lot. As he was walking, he looked for "numerical and geometrical patterns—in the sidewalk, in house design, tree shapes (fractals came natural to me in later life)" (personal communication, November 11, 2015). Even in his early eighties, he would look at house shapes through the eye of patterns when he was able to walk with his dog, Miss Buffy. As he later found out, there are not only regular patterns but also chaotic patterns. David Kirshner (in this volume), a friend and colleague, mentions Doll's own metaphor of the chaotic path he followed during the period between his divorce and his second marriage. While the curve was chaotic, he was able to sustain the tension at the boundary of order and chaos and emerge from it with a more sophisticated understanding of life and teaching.

Fascinated by the patterns of nature and life since childhood, William Doll has introduced this patterned view into teaching. At Shady Hill School in Cambridge, where Doll worked with professors, teachers, and students, everyone played with mathematics on Friday afternoons. They frequently played with Cuisenaire Rods, which are little square rods in different colors and lengths: one, two, or three centimeters. Playing with different combinations of rods led students to "see" patterns of mathematical thinking. They also went beyond whole numbers: "What happens when we call the white rod 1.5? or 1.3?" Playing with pattern after pattern with students helped Doll understand that "mathematics is a sense of patterned relationships" (interview with Doll, 2009).

Mary Aswell Doll also recalls how William Doll used Cuisenaire Rods in a parent–teacher conference at Valley School in Baltimore "where Bill demonstrated for a group of parents how the rods worked, and how students could manipulate objects, Maria Montessori–fashion, to obtain a physical understanding of mental processes" (personal communication, December 30, 2015). It is important to acknowledge how Montessori's notion of playing with objects, in addition to Piaget's theory, influenced Doll's early work.

Patterned relationships become a key component in Doll's pedagogy. He has developed a pedagogical approach to working with patterns in multiple layers. First, each subject has its inherent patterned relationship. While mathematics is a great example, other subjects also have their internal patterns. Doll emphasizes the need for teachers to know their subjects well, and he does not mean the facts or information, but the structure—patterned relationships—of the subject. Only by getting in touch with the fundamental patterns of a subject can a teacher be free to think in more depth and guide students to explore ideas more thoroughly.

Second, students' thinking has its own patterns and the bridge between the patterns of a subject and a student's thinking is play. The human mind has a curiosity for patterns and when students are guided to "see" patterns through playing with them, they learn the most. Third, patterning is also emergent and nonlinear. Curriculum and pedagogical structure emerges through the interactions in the classroom and follows recursive, circular, web-like movements rather than linear lines. A pedagogy of patterns forms and transforms itself. Fourth, influenced by Gregory Bateson (1979/1988), the "pattern which connects" (p. 8) is both ecological and spiritual, as it connects not only human but also all living creatures. Such an "ecology of mind" (Bateson, 1972) leads to awe, wonder, and humility toward the interconnectedness of life that is ever shifting (Doll, 2011b). Doll's pedagogy of patterns emerges from a nonlinear patterning of his life, thought, and teaching, which is both recursive and open.

Teaching at School Through a Sense of Patterning

> The device we use for developing the student's structures, particularly those structures relating to arithmetic, is a sense of patterning. The whole curriculum-teaching part of the [Oswego-Sotus] Project is built around patterning: copying patterns, building patterns, extending patterns, and finally transferring and transforming patterns. (Doll, 1984/2012, p. 78)

Acknowledging that students' thinking and subject matter both have structures, William Doll (1984/2012) developed teaching devices for encouraging students to think through "a sense of patterning" (p. 78). To be able to discern the patterns of a subject matter requires imaginative insights beyond facts and commonly accepted rules. Through his own experiences as a schoolboy, a teacher, and a parent of a schoolboy, Doll believes that "at all levels, from kindergarten through graduate school, mathematics can be dealt with meaningfully as 'playing with patterns'" (Doll, 1993, p. 176). Mathematics taught as memorizing games drains the spirit from the subject—both the school subject and the human subject (Pinar, 2014). Presenting mathematics as patterned relationships, Doll has found ways of encouraging schoolchildren to get in touch with the vitality of mathematics. For example,

> When I was teaching elementary school (back in the 1960s) my fourth grade students knew their multiplication tables at least through 20 × 20. They did not, though, memorize 400 separate facts; instead they found patterns which they played with and abstracted: 6 × 17 is double 3 × 17, itself 30 plus 21. . . . All this led to a playful/abstractive sense of working the numeration system in countless ways. (Doll, 2008/2012, pp. 28–29)

Doubling is a device Doll often uses to help children see mathematical patterns, and the flexible and countless ways of doubling to add, subtract, multiply, and divide numbers brought a lot of fun discoveries to children who intuitively played with patterns. One of Doll's favorite quotes from Whitehead is, "What you do teach, teach thoroughly. . . . Let the main ideas which are introduced to a child's education be few and important, and let them be thrown into every combination possible" (Whitehead, 1929/1967, p. 2). When children have the opportunity to throw things into every combination possible, they begin to see connections, relationships, and patterns that they had not seen earlier.

Another example Doll likes to give is to compare the following two charts:

Chart 4.1: Common number sheet

1	2	3	4	5	6	7	8	9	10
11	12	13	14	15	16	17	18	19	20
21	22	23	24	25	26	27	28	29	30
31	32	33	34	35	36	37	38	39	40
41	42	43	44	45	46	47	48	49	50
51	52	53	54	55	56	57	58	59	60
61	62	63	64	65	66	67	68	69	70
71	72	73	73	74	75	76	77	78	79
81	82	83	84	85	86	87	88	89	90
91	92	93	94	95	96	97	98	99	100

Chart 4.2: Doll's number sheet

0	1	2	3	4	5	6	7	8	9
10	11	12	13	14	15	16	17	18	19
20	21	22	23	24	25	26	27	28	29
30	31	32	33	34	35	36	37	38	39
40	41	42	43	44	45	46	47	48	49
50	51	52	53	54	55	56	57	58	59
60	61	62	63	64	65	66	67	68	69
70	71	72	73	74	75	76	77	78	79
80	81	82	83	84	85	86	87	88	89
90	91	92	93	94	95	96	97	98	99

In Chart 4.1, the relationship between numbers is shown but is not fully evident; in the Chart 4.2, the relationship among numbers is visually clear.

Chart 4.1 neglects the important role of 0, not even including it. However, the inclusion of 0 in Chart 4.2 makes the patterns of numbers come alive. Doll understands that 0—a commonly neglected number because it is considered "nothing"—plays a central role in the numeration system. When children acknowledge the role of 0, they can understand the meanings of negative numbers—without using the term—much earlier than what is commonly assumed. To go beyond commonly accepted assumptions, the teacher must play with mathematics unconventionally. The teacher's passion for and insights into a subject matter form an important entrance into a pedagogy of patterns.

Since Doll became a teacher educator, he has continued to play with the number line at school—in various forms—either with pre-service teachers or with colleagues. He considers the number line a crucial building block for developing children's mathematical thinking and capacity for playing with the "pattern that connects" (Bateson, 1972; Doll, 2012; Reynolds, 2005). When teaching at LSU, he brought pre-service teachers to schools to teach mathematics using the number line. Going up and down the number line, students gained insights into the structure of the number system and developed their structural mathematical abilities. Doll also participated in students' play while supervising student teachers. With David Kirshner, Doll once team taught a group of fifth graders to explore the meaning of fractions through a ratio, which ended up with a new category: a fraction of a fraction (Krishner, in this volume). Playing with patterns advances students' thinking.

At SUNY Oswego, with a graduate student, Doll established an arithmetic project for schoolchildren to play mathematically:

> I established relationships with students and did projects with them. One of my students came in from Sotus, New York, and we established the Oswego-Sotus Structural Arithmetic Project and did projects with school students. I brought with me all sorts of stuff I used as a teacher in Boston. We played with symbols and looked at patterns. Sometimes we played the sequence: 2, 4, 6, what's next? [The number] 10 occurs only if you say "if we continue the pattern." What is important is that a kid begins to see a sense of logic emerging and sees clearly what the rule is: "if, then." We played to see patterns and understand the assumptions of how base 10 is founded. A whole new world emerges out of base 2. We played with different bases. The project went on for three years. . . .
>
> The project was intended to teach kids to think for themselves. One of the devices I used was to turn the number around. For a very simple example: number sheets, 10 by 10.

They have addition, subtraction, multiplication, and division sheets. For the third or second graders, to subtract 3 from 0, we got a negative number. [The number] 4 divided by 7, for example, was a challenge. Our kids were quite prepared to work with unusual stuff, and they were not afraid to handle the unusual. For other kids, they said, "We cannot do it." For our students, they had no problem dealing with that which they had not seen. They had a lot of time to see the unusual and play with it. I did it with [graduate student] Jim Wood together. I developed the design. (Interview with Doll, 2010)

Schoolchildren enjoyed playing with patterns and grew more confident in dealing with the unconventional. Conventions, which are actually not natural to their curious minds, are usually imposed on schoolchildren. In Doll's pedagogy of patterns, he believed early on in his teaching career that children have the ability to see the world in a patterned way. As I explain later, Chomsky's theory of linguistic development, Bruner's theory of cognitive structure, and Piaget's theory of knowing through construction of schema all supported Doll's assumptions in his experiential discovery. His readings of these thinkers became intertwined with his own explorations and adventures.

Patterns of subject matter must be mediated by students' own patterns of knowing. Patterned relationships emerge through exploration and cannot be directly delivered to students. Introducing play into teaching provides an important bridge for students to think, to explore, and to adventure into the unusual. The play that Doll implemented in the classroom is open-ended, without rigid, clearly laid out rules of what is right or wrong. It opened up possibilities of reorganizing patterns in creative ways.

This openness caused Doll to embrace nonlinear teaching. The patterns of a subject are fixed but can be revealed through multiple lenses. Students do not think in a linear way but follow curved and recursive pathways. Only through their playing with different combinations can the intricacy of the web-like reality of each subject be revealed. While this nonlinear patterning was already embedded in Doll's early teaching, it became an explicit focus later in teacher education, to which I turn next.

Playing With Patterns in Teacher Education

Mathematics-oriented chaos theory with its emphasis on nonlinear recursions and physics/biology oriented complexity theory with their emphases on structural transformations present an open frame which encourages participants, players, observers to think in terms of, and work with, patterns, networks, webs. Again, here lies the heart of a new, post-modern pedagogical paradigm. (Doll, 2011d/2012, p. 159)

Competency-based education was influential in teacher education in the 1970s. It refers to "a teacher training program in which there are specific competencies to be acquired, with corresponding explicit criteria for assessing these competencies" (Bowles, 1973, p. 510). Such criteria include knowledge, performance, and product (Arends, Masla, & Weber, 1971). Making a distinction between performance and competency, Doll (1977, 1984/2012) argues that cultivating competence in teacher education is important but we need an alternative model not restricted by behaviorist objectives and measurable outcomes. Unfortunately, competency-based education privileges performance and measurable results.

Doll (1984/2012) proposes a competency model that values the "structural patterning" (p. 68) in Chomsky's linguistic competence, the transformation of cognitive structure in Piaget's equilibration process, and competence motivation in Bruner's theory. In this model, students' competence is developed through a sense of patterning. The Oswego-Sodus Structural Arithmetic Project (for specific examples, see Doll, 1984/2012, pp. 78–80) is a good example of the implementation of this model. This sense of patterning also exists in other subjects, including in both natural and social sciences.

After William Doll's encounter with postmodernism and chaos and complexity theory in the 1980s, the structuralist orientation of patterning began to give way to a more complex viewpoint of patterning in which emergence and self-organization play an important role. His further exploration of ecology and cosmology has led him to take a broader view of social, cultural, and cosmological connections in which his pedagogy of patterns are situated. To shift one's focus from knowledge or information or performance to patterns that connect is an ontological transformation in which relational knowing, situated in intellectual, cultural, and spiritual connections, becomes the foreground of the pedagogical landscape. In this landscape, the creative emergence of new ideas and structures comes from relational interplay rather than individual components, in which change is no longer accumulative but transformative (Doll, 1993, 2012) with a *gestalt* switch.

Fractal Entrance to Patterns

Fractal images are among Doll's favorite metaphors for reconceptualizing curriculum and teaching. Fractal geometry combines nonlinear mathematics, computer programming, and aesthetic presentation to capture the fluid and irregular complexity of reality, including complicated shapes in nature such as leaves, coastal lines, cloud formations, and ocean waves. Fractal images in the video *The Mandelbrot and Julia Sets* (Art Matrix, 1990) show an endless sequence of patterns within patterns that are not identical but similar across different scales. Doll often uses this video in teacher education classes and asks students to examine nonlinear movements and their implications for education.

The entrance into fractals is not singular or straightforward but multiple and circular, because the interplay between the part and the whole in fractal images makes it impossible to see a clear beginning, or an ending, in a view of patterns embedded in patterns. We can follow the curves of different segments of fractals to dive into more details while those details as unfolded embody the whole. Both structure and surprise—two essential elements of self-organization in complexity theory—are demonstrated simultaneously. Different entrances lead to different paths. As James Gleick (1987) points out, "Every foray into the Mandelbrot set" brings "new surprises" (p. 228).

Doll adopts a fractal entrance to teaching. He often sends out teaching packages or notes ahead of time, before the class starts. In those notes, he might make an interesting statement to unsettle commonsense assumptions and draw students' attention to another way of looking at an issue. He might ask provocative leading questions to provide students an opening into rich material through multiple routes. Or he might write a short essay regarding the reading materials to invite students' responses. All these simple yet rich beginnings encourage students to enter the class with curiosity, thoughtfulness, and a critical eye.

Petra Munro Hendry commented on William Doll's simple teaching device: "He just throws a statement out or he'll pick one line from the text. Then ask: 'what do you think?'" (interview with Hendry, 2010). But such a simple question could open up a most intriguing conversation. The simple questions he asks are often not quite as simple as they may at first appear, nor are they random, but serve as openings to the complex patterns of a fractal curriculum structure. Jie Yu, a former student, commented on a class taught by Doll:

> Rather than giving the class a rigid syllabus at the beginning of the semester, Dr. Doll threw a seed full of complexity to us and then worked patiently with each individual student to develop different things from that same seed. (Personal communication with Yu, October 31, 2010)

Doll (1999/2012) uses the metaphor of planting a seed, from both agriculture and chaos mathematics, "a seed rich in problematics or possible interpretations/analyses/procedures" (p. 215), as the beginning point for weaving the web of teaching and learning. In such teaching, dialogic relationships, multiple perspectives, the formulation and reformulation of patterns, and interrelated connections within the subject matter are in a continual process of emergence. Here entrance into the pre-set structure of a discipline in linear teaching gives way to fractal entrance to complex emergent patterns in nonlinear teaching.

Fractal entrance to the pattern that leads to holistic, matrixical teaching is enabled and enhanced by recursive practice. Hendry recalled, "Bill had that notion of recursion and we practiced it several times in the class. We went back and we read and re-read and then we would go back again. We cycled through texts multiple times" (interview with Hendry, 2010). The focus was never on "mastery" of the text but on drawing on text to generate more questions. Typically, Doll leads the class to make all sorts of connections between and among different authors or texts and asks student to relate to readings through their own experiences and reflective thinking. He may pick up what the class is reading in week five and connect it with week two or three, or find a particular recurring theme and deliberate on it each time in a somewhat different form. He also sends notes to students throughout the course by email to make additional comments, so there is constant feedback to form a network of relations in which the entrance becomes recursive.

Thus patterns of curriculum emerge out of all the interactions and students' engagement with texts and with others. Complexity and layers of meanings are built on one another while students' engagements with knowledge and with the instructor are also deepened. In other words, patterns are nonlinearly emergent. As Sarah Smitherman Pratt (2011) points out, "When new ideas emerge out of conversations, research, and experiences, these ideas are a result of the system of interactions and also fold back into the system to become part of the ongoing conversation" (p. 43). In this way, the system is in an ongoing process of transformation, as are all participants in the system.

Fractal entrance to patterns can best be accomplished through play. Doll vividly describes a math game:

> I was teaching a math disciplines course and we were struggling with fractions for a 3000-level class. David [Kirshner] came in and he said, "I've got a game," and he produced a wonderful game. The elements of the game were simple: a string, some paper clips, and some rectangular cards, and the cards with fractions on them. He broke things down to do simple and complex fractions. A simple one would be 3/8 and a complex one would be 1/2 of 3/8. You gathered these cards (one side was blank

and the other side was the fraction) and let a student pick a card, let's say the card is 3/4, and you asked students to put it on the line (string). That was, of course, mind-boggling. And the class chipped in, "Put it here," "Put it there." We calmed the class down, and I said, "Where do you put it?" The student tried and asked: "Is it okay?" I say, "It is up to you."

Another student picked up a card, and it might be 2½. The student said, after a bit of hesitation, "Well, I could put it on the right of 3/4 or on the left of 3/4. Could I put it anywhere on the line?" I said: "It is up to you." The student sat down and now was looking at the two places where the cards were put. The next student picked up a card, and let's say, 1/2. Now the world changed because 2½ and 3/4 had set a certain parameter. The 1/2 could not go between those two and could not go to the right of the one but has to go to the left of 3/4, so where? You had to figure out where to put 1/2. It was a simple, simple game, but it dealt with relationships. David and I had a wonderful conversation about fractions. (Interview with Doll, 2010)

I observed this game in that course when I was Doll's teaching assistant. I have to admit that at the time I was not particularly impressed because it was simple math. But by understanding it from a pedagogical viewpoint, the game leads pre-service teachers to the formation of patterns and also provides a great tool for them to use in their own teaching. The entrance to the number line is not predetermined but the interaction of multiple entrances leads to mathematical relationships. The complexity of the number system, including the role of zero and negative numbers that are often not introduced until upper-level classes, can be introduced to students in the early grades through play. I also visited schools with Doll and his undergraduate students—pre-service teachers—in that course. It was amazing to watch how the first graders played with the number line to understand mathematical patterns.

Complex Patterns and Nonlinear Teaching

Since Doll started his career in teacher education, he has struggled with the question of how to help his students develop their own teaching styles that emphasize a sense of exploration and play. Over the years he has developed a mode of nonlinear teaching in which fractal patterns that represent "both change and stability" (Reynolds, 2005, p. 266), "frame and flexibility, [and] structure and spontaneity" (Doll, 1999/2012, p. 209) emerge in a self-organizing classroom. Self-organization refers to a system's spontaneous reorganization when a critical point far from equilibrium is reached (Prigogine & Stengers, 1984). In other words, a new order emerges near the edge of chaos

(Kauffman, 1995; Prigogine & Stengers, 1984). Complexity defined in this context emerges from "interactive, dynamic systems that under *specific and limited* conditions are able to transform themselves" as a result of interactions among local components (Doll & Trueit, 2010, in Doll, 2012, p. 175; italics in the original). For decades, Doll has explored how to create pedagogical conditions for a classroom—a social system—to promote students' self-organizing capacities.

Drawing on Prigogine's self-organization theory and Stuart Kauffman's complexity theory, Doll (1989/2012, 1998b/2012, 2008) discusses the pedagogical implications of emergent complex patterns in a self-organizing process and proposes—and practices—several important conditions: Flexibility rather than a predetermined plan must be built into teaching strategies and lesson plans to allow emergence; a certain degree of perturbation—not imposed but willingly encountered—is necessary for initiating transformative dynamics; and adequate time must be provided for students to explore, play, and interact with patterns. It is a systemic view of teaching that emphasizes the importance of interactions to make transformative learning possible. While pedagogy of perturbation is discussed in Chapter 2, here I focus on the mixture of stability and flexibility, the mutual embeddedness of complexity and simplicity, and the role of interaction situated in temporality to discuss a pedagogy of complex patterns.

The Mixture of Stability and Flexibility

The mixture of structure/stability and surprise/flexibility is an important condition for self-organization to happen (Doll, 1998b; Kauffman, 1995). Here the sense of stability is not static but fluid and interdependent with changes that result from interactions of local components. The sense of flexibility does not fall into the abyss of chaos, but stays at the edge; tension between the two is essential for the emergence of a new pattern. Emergent curriculum and pedagogical structure indicates both a sense of boundary and unpredictable specific pathways, as the term "strange attractor" in chaos theory suggests. A strange attractor indicates that a chaotic system is attracted to a certain area in which a regular pattern can be detected but the specific motion of a particular pathway cannot be predicted (Fleener, 2002). Such a combination of boundary and uncertainty can be seen in complex class interactions.

The following is an example of emergent patterns incorporating both stability and flexibility in the weekend class Doll and Trueit taught in the spring of 2013 (see Chapter 2). In one of the sessions when Doll was the sole instructor, the class had just discussed the historical and comparative concepts of the child in Tröhler's (2011) book *Languages of Education*. While Tröhler focuses on a historical reexamining of Rousseau and his theory, a Chinese visiting scholar in the class who was an early childhood educator had commented in class and talked with Doll on various occasions about Fröbel, Pestalozzi, Rousseau, and their intellectual relationships. Doll wanted to illustrate these concepts so he gave the class a break. During the break, he and his teaching assistant put poster paper on the wall for him to write on. As the students returned to class, he asked for several minutes outside of the classroom in order to concentrate on his notes.

After William Doll synthesized the class discussions of the book in his mind, he came back and wrote on the poster paper a diagram while explaining it to the students:

Fröbel

Pestalozzi Rousseau

Child development

(and Tröhler)

       History (what is H?)   Romanticism

R (through Tröhler) ⟵⟶ F

Citizen (growth and progress)   Soul, spirit (unity, mysticism, purity)

Child development

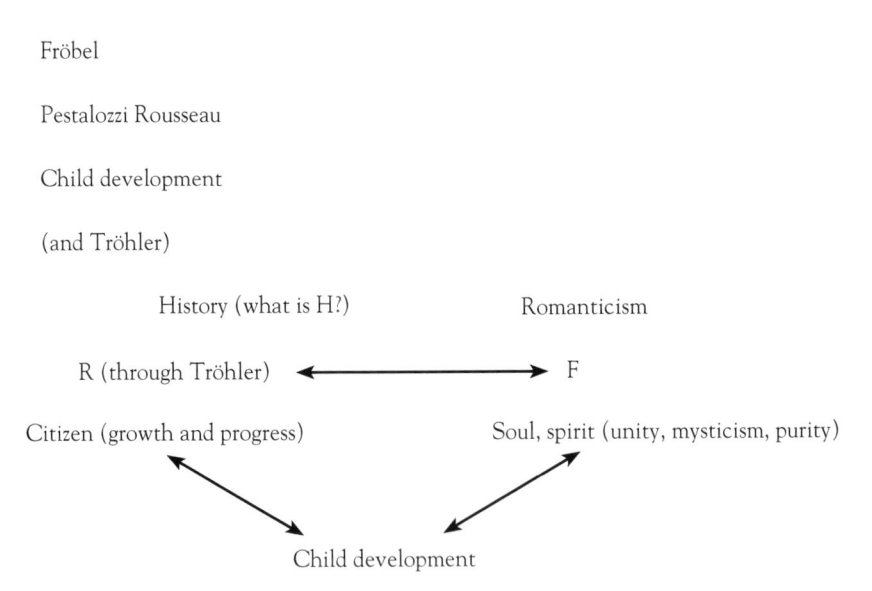

Figure 4.1 An Emergent Diagram in Teaching

This diagram had emerged from previous class discussions. What Doll wrote on the poster paper was a relationship chart of important concepts from different thinkers. He explained that context is important to understanding these relationships and that there is a difference between the United States and Germany regarding the notion of the child. A new way of approaching the child became a measurement approach for Americans. The American idea of measurement took over child development without paying attention to the pedagogy of the child, which was part of the German notion. As he created and explained this chart, the class continued to discuss related ideas (from video transcription, March 23, 2013).

If we see this diagram as a strange attractor, then the class discussion had gone around it but each student's specific understanding was unpredictably different. The goal of teaching for Doll is not to control the outcome but to open up space for students to form their own understandings. Here his diagram incorporated students' input that was beyond the scope of the book. The one required text provided the boundary for the class discussion, and the diagram illustrates an emergent pattern that integrated both structure and surprise.

In nonlinear teaching, different from predetermined linear teaching, the teacher does not decide everything in advance but gives students a chance to explore and discuss among themselves, during which diverse perspectives and multiple possibilities emerge. The teacher does not respond to every student's comment to provide a definite answer, but picks up some ideas for the class to explore further. Over time, dialogical interactions lead to a matrix of relationships between and among teacher, student, text, and context. One thing that Doll consistently does in teacher education is hermeneutical tracing of important concepts. Such an inquiry unfolding in the classroom often becomes interdisciplinary during the process because students come from diverse intellectual backgrounds. Following is another teaching scenario from the same class.

Doll:	Now then, you start off and you talk about physical literacy coming from the UK, is that true?
Pete:	Yes.
Doll:	The word "literacy" seems to be popping up in a number of areas and as a reader I am curious about why.
Pete:	UNESCO made 2003–2013 a decade of literacy in education. That's why everybody has been so taken by it.
Doll:	Can you give us some history of that and any analysis of why they want to use the word "literacy"?
Pete:	I don't know, but I just read an article about the Canadian position on it, its background, and UNESCO's use of "literacy."
Doll:	It may or may not be important for your paper, but as a PhD student, one of the things that you can do . . . well, it might be difficult to track down why UNESCO picked up this word, and it is not necessary for this paper. But you certainly can go to the OED to look at that word and what it means. I find the notion of scientific literacy, physical literacy, et cetera, not a combination I would put together. "Literacy"

Guanglu: is catchy, but I am not certain why they pick up that word. I've found it unusual at most.

Guanglu: I think it goes back to Hirsch. It is a conservative term. It means, for example, if you are an immigrant to America, you'd better give up your mother language and culture and accept the American standard of knowledge, and then gradually become an American and adapt to American life.

Pete: Yeah . . . I talk about it from another way. Whitehead talks about what defines physical literacy and in the middle of her paper she mentions competency. I use competency and pleasure together since pleasure is missing from it. In the first paper she wrote in 1990 and ever since she warns against putting pleasure in it.

Doll: Oh, good that you can bring it back. It is always good!

Pete: And then, instead of physical competency—since it brings back judging the body and there is a danger in trying to measure physical competency as if it is something that can be achieved—she means it in a phenomenological way. Then I can go to Aoki and discuss pedagogical content, being able to read the environment.

Guanglu: For this word, first we must set up a standard, such as the middle-class standard: middle-class vocabulary, middle-class knowledge. This word is popular in China.

Doll: Really?

Guanglu: Yeah! Computer literacy, writing literacy, mathematics literacy, et cetera. It is a popular word in China, although there are a lot of critiques. Generally speaking, it is conservative as it originates from E. D. Hirsch.

Pete: It is functional, right? Okay. If we look at Canada, a cosmopolitan place, in my classroom, and all over the world, and I want to draw upon Aoki to talk about a particular community of difference and envision something different. In Britain and Europe they talk about using literacy to embrace difference. "Reading" is used in a different way to include such as reading people, reading each other, and reading the environment. Reading landscape is the big thing in the BC, and we can talk about being literate from an indigenous way as reading the land and listening to the land. Children are in the environment and part of the landscape and then the question is what physical education can do with it. That is where I want to go, my different spin on it to make it connect with relationships, rather than just a functional and instrumental sense.

Guanglu: For example, reading literacy, it means that you have to master vocabulary that every child must learn.

Pete: So, if we have immigrants come to my class and do physical activities, I don't think they have to do it in a Canadian way. If we tried to embrace activities from their own cultures, it would be different. I

	want to do this kind of work, and it is kind of big and difficult to write. Hopefully I can try to capture it.
Doll:	One can make the distinction between "literate" and "literacy," and "to be literate," and again I will need to go to the dictionary, but I think it would mean to be competent, to understand, to know a field or a subject, and I can see E. D. Hirsch uses it as a political tool to make sure that all students understand a standard, understand in a particular way, and one can go further, understand the American way as the right way. Fascinating!
Guanglu:	In his opinion, education is about reproduction, not creation.
Doll:	Yes.
Guanglu:	Many Chinese scholars use this word. After I do some readings, I think we probably forget the original meaning of this word.
Rasunah:	The original use of "illiterate" means that you are not able to read and write. Actually, the standard is not a high standard with literacy. So when you talk about world literacy, and you have a mixed class, how would you be able to describe that result? You should be able to do it now with classes. You talk about literacy as the microcosm of what is possible.
Pete:	Yeah, I talk about it as relational, so it is about getting to know each other, reading others and reading the self.
Rasunah:	Can you find a word more dynamic? Or has more substance? Because you can change that and you don't have to stick with that word.
Doll:	I would think to get published, to catch people's attention, one needs to be willing to use it. But as we have talked around, one can play with that word, maybe in this paper, maybe not, the notion of literate, illiterate, and raise some of the issues being raised. I think that is important.
Pete:	Merleau-Ponty's foundation is phenomenological, and physical literacy influenced by it is a continuous process rather than an aim to be achieved. Language can be confusing, so . . . .
Rasunah:	I understand that in Canada, in the teaching profession, originally literacy was perceived as remedial, and it is just fashionable to use the word "literacy" . . . .

[Rasunah and Pete engaged in a prolonged conversation about the relationship between the remedial and literacy, and then the whole class shared experiences of study or teaching related to remedial classes.] (Video transcription, March 23, 2013)

From this class conversation, we can see that Doll enacted a hermeneutical approach to understanding the notion of literacy. As the discussion evolved, different concepts of literacy were presented, including Pete's effort to develop a relational and phenomenological approach to embracing difference when formulating physical literacy; Guanglu's critique of a conservative, political approach to literacy as the standard; and Rasunah's association of literacy with remedial education. Students were also challenged to go beyond their comfort zones in participating in this conversation. Sometimes Guanglu's response followed his own thinking rather than intersecting directly with his classmates' comments, but over time all different approaches had adequate space and time to be discussed. Divergent thinking was encouraged and there was no pressure to reach a consensus or closure.

Doll was not bothered by the "excess" or overflow of information, but allowed the juxtaposition of ideas to unfold. As Janet Miller (2005) points out, "excess" points us to what cannot be contained and highlights the role of differences for working out meaningful relationships: "The strategy of juxtaposition is one that invites inconsistences, ambiguities, ambivalence, and . . . 'unspoken themes'" (p. 114). Difference is not simply left alone because "working difference" (Miller, 2005) and playing with difference (Doll, 2012) require interactions, although convergence is not demanded. Doll did not respond to each student's comment in a linear mode, but picked and chose from students' diverse responses to invite further discussion, a good example of emergent patterns in nonlinear teaching. To use Doll's own words:

> The field or area of study in its multiple facets, various perspectives, and interrelated connections begins to emerge. While a student, or students working together, may well pursue particular aspects of a subject, through interactions all begin to interconnect their own particulars with those of others. In this manner a holistic image emerges. Whereas the linear mode is often associated with a reductionist methodology, a non-linear mode is associated with a holistic, matrixical methodology. (Doll, 1999/2012, p. 215)

The Mutual Embeddedness of Complexity and Simplicity

It is important to point out that complexity and simplicity are mutually embedded. One of the conditions for self-organization—the linchpin of thinking and teaching complexly—to happen is this simple rule (Doll, 1998b; Kauffman, 1995). As Fritjof Capra (1996) points out, in the nonlinear world simple rules may lead to richness and variety while "complex and seemingly chaotic behavior can give rise to ordered structures, to subtle and beautiful patterns" (p. 123). Nonlinearity intertwines simplicity and complexity in unpredictable ways.

I have discussed Doll's practices of throwing a few major ideas into as many combinations as possible at school. In teacher education, Doll has seldom used more than three texts at the graduate level whereas other professors usually assign many texts. He guides the class in conversation about the major ideas of each text in the context of each student's own experiences. He also asks students to select their own readings and discuss them with the class as a whole. In this way, the main ideas of required texts are thrown into all sorts of different combinations with students' own readings and interpretations.

In Doll's method, fewer texts do not reduce the richness or complexity of class conversations, because these writings are more fully discussed through interplay with students' (intellectual and social) experiences and critical reflections. Multiple layers of understanding emerge from class discussions. The class becomes more meaningful to students because they have substantial input into curriculum design, and they are part of the complex, recursive cocreation of pedagogical structure. When possible, Doll takes a further step, giving up required readings and letting the class readings emerge more fully from each student's own choices. Jie Yu comments on a team teaching class co-taught by Doll and Hendry:

> The "Chaos, Spirituality and Education" [class] co-taught by Dr. Doll and Dr. Hendry is one of the most impressive classes I have ever taken in our department. From the very beginning of that class, I recognized its uniqueness because I did not get a syllabus specifying what readings I had to finish on which dates, as I always got from other classes. As both Dr. Doll and Dr. Hendry would recommend particular books for us to read, we were not assigned fixed schedules of reading. We discussed our research interests together in class and then made our own decisions of what we would like to read for the following week(s). The plans were always emergent. And everyone in the class finally had a different reading list, working at different paces. (Personal communication with Yu, October 31, 2010)

In such a pedagogical design, the shared texts are recommended rather than required so the provided structure is changeable. Here Doll's pedagogy of patterns connects professors, students, and the subject of teaching and learning in a fractal and nonlinear mode. The complexity of the learning and teaching happening in the class is made possible by the simple change of not requiring readings but allowing reading lists for each individual student to emerge. The team teaching model also adds a layer of complexity, as I discuss throughout the book.

The simple rule of choice did not mean it was easy to implement, as it must have been confusing for students who were accustomed to having a clear-cut syllabus with assigned readings. As confusion subsided, however, students experienced the intellectual freedom that is often made available to professors rather than students. Students' involvement in learning became an essential component of pedagogical design. The most complicated process of learning can happen with a simple change in the initial pedagogical arrangement, but it is a change that has a profound impact on the pedagogical orientation and interactions that make "a complex conversation" possible (Pratt, 2011, p. 44). Nonlinearity is inherent in this design by weaving layers upon layers of meanings recursively and interactively.

Interactionism and Temporality

Doll argues that "thinking complexly means looking at the situation and the interaction of all participants in that situation" (interview with Doll, 2010). For Doll, the centering of individual entity, whether it is an individual person, or a text, or society—child-centered, teacher-centered, subject-centered, and society-centered curriculum theories have historically been debated to gain predominance in the field (Marsh & Willis, 2007)—does not capture the complexity of education. It is not any individual component in a system but the interaction among all components in the situation that fertilizes moments of self-organization.

Such a privileging of interaction requires a systemic approach in which relationships are not static but emergent (Broom, 2011; Doll, 2011b). According to Sam Crowell and David Reid-Marr (2010), "What emerges goes in all directions at once, influencing everything and being influenced in return" (p. 118). In other words, the mutual and simultaneous influence of the local and the global as well as interactions between local components lead to new patterns. In this sense, creativity becomes co-creative (Wang, 2010) because emergence is impossible without the relational.

Centering emergent relationships in teaching also means that a teacher needs to allow the process to unfold by giving time and space for complex and nonlinear interactions to happen. Sometimes it does take time to let students explore and play with the patterns of subject matter, patterns within their own thinking, and patterns of others' thinking. The foregoing video transcription of teaching demonstrates the willingness on the part of the pedagogical authority to allow students to take time to engage in divergent discussions, in which cross-fertilization of different ideas can occur. Students left the class without reaching agreement about the notion of literacy but their understandings had been enriched. As Dwayne Huebner (1975/2000) points out, temporality is not a characteristic of isolated individuals but is instead "a characteristic of being-in-the-world" (p. 244) and exists in encounters between the individual and other individuals and with the world. Experiencing such encounters temporally, students expand their horizons.

A doctoral student from South Korea, Jeong Suk Pang (in Doll, 1999/2012) also commented on the role of time in her observation of Doll's teaching. The confusion, chaos, and uneasiness students experienced when they first encountered Doll's unconventional teaching gradually gave way to their active participation in class interactions and thoughtful crafting of their own teaching styles. As Doll (1993) points out, in a postmodern, complex framework, "time takes on a different, qualitative dimension; it acquires a transformative aspect, since development of one sort or another is always occurring" (p. 179). Without allowing the temporal dimension of learning to play out its crucial role, pedagogical potentiality is lost.

Most examples that Doll gives regarding this new pedagogical paradigm are about mathematics teaching because of his experiences and the nature of the discipline (see Doll, 2012, Chapters 2, 3, 7, 17, 20, 22, and 23). However, Doll also contextualizes mathematics teaching in the social and political context so that mathematical patterns are connected with cultural patterns. In his formulation of the 4 R's, "relation" includes both pedagogical relation and cultural relation and these two are intertwined (Doll, 1993). He regularly taught an undergraduate mathematics discipline course to pre-service teachers at LSU and chose *Flatland* (Abbot, 1884/1992), a novel, as one of the required texts for this class.

A Square is the narrator of this intriguing novel that uses geometric figures to portray different worlds in different dimensions including a two-dimensional flatland, a three-dimensional world, and, in the imagination, a four-dimensional world. It is full of mathematical imagination, but its underlying message is a social and political critique of the hierarchical system and culture. It challenges readers to go beyond the limitations of their own dimension. Students in Doll's class discussed it and wrote short-response essays. They were fascinated not only by the mathematical concepts and patterns but also by the novel's description of gender roles and social systems using mathematical language. The bridges between the mathematical and the social that had been mostly invisible for them now became clearer.

> *Autumn 2000. I am accompanying Dr. Doll and Donna Trueit on their academic tour in China. We have had a chance to tour some gardens, going from northern China to the most famous garden cities in Suzhou and Hangzhou in the south. Dr. Doll maintains his excitement with everything he encounters, from the pattern of pathways that are made from small, irregular stones to the similar yet different designs of windows in the gardens. He keeps calling "fractal!" to our attention as he looks at things with a fresh eye. I have never considered something I have been accustomed to for so long as fractal. But a stranger's eye helps me see*

fractal patterns (Wang, 2005). *In my further studies, I've found that fractal principles, not by accident, have been evident in Chinese paintings and garden design, which are connected to the overall Chinese philosophy and culture, which is more interconnected and matrixical. Unfortunately, Chinese political and educational systems have remained hierarchical, as Dr. Doll notices right away when he walks into a Chinese classroom. Can the complexity of the cultural relations in Chinese aesthetics infuse life into the pedagogical relation and move it away from hierarchical control? Contemporary Chinese curriculum and educational reform is making such an effort to vitalize school education (Pinar, 2014), and Dr. Doll has made his contribution to this effort for the past decade through his work with Chinese educators, including implementing the 4 R's in teaching Chinese language and math at school. It remains to be seen to what degree cultural transformation may happen simultaneously to support the educational movement toward more freedom and space for exploration.*

Kauffman (1995) discusses the significance of self-organization not only in nature and science but also in society, life, and the cosmos. The cosmological implications of self-organization appeal to Doll's sense of spirituality and his effort to build an open curriculum system that integrates science, story, and spirit. Doll's approach to the pedagogy of patterns is also ecological in the Batesonian sense, in which relationality lies at the heart of life and education. Simultaneously intellectual, cultural, spiritual, and ecological, a pedagogy of patterns, located at the center of this book, connects the 7 P's of William Doll's pedagogy.

· 5 ·

PEDAGOGY OF PASSION

> I see our task as curricularists and instructionists to look at the complexity science with an eye to seeing the spirit inherent within these sciences, a dynamic spirit featuring the interplay of *passion and play*. (Doll, 2003/2012, p. 110; italics in the original)

William Doll is passionate about teaching and teaches passionately. It is obvious to anybody who experiences his enthusiasm for students' learning and growth. When I was his graduate assistant at LSU, he always came out of class with a hyperbolic gesture: "What a great class!" I also remember how joyful he was around the children when he took student teachers to schools and played with mathematical patterns using the number line. After he retired from LSU, he moved to Victoria, Canada, in 2007, and now teaches at the University of Victoria and the University of British Columbia as an adjunct professor. Teaching has been such an intimate part of his life that it seems to be impossible for him to be fully separated from it. In this chapter, I discuss his pedagogy of passion that links his joyful spirit as a person, his devotion to students as a teacher educator, his enthusiastic play with intellectual ideas as a scholar, and his eagerness to learn from others in a distant country.

Joyful Commitment to Students

> Being with the same students over a number of years was a joy and by now was a *modus operandi* for my work in the schools. (Doll, 2009/2012, p. 17)

Doll is committed to students' growth and takes pride and joy in their progress. As Stephen Triche points out, William Doll "takes joy in his students; that is probably the greatest thing I have learned from him. [That is] to show the joy I have in students [in my own teaching], being joyful around them, even when there are a lot of problems" (interview with Triche, 2012). Doll is a joyful teacher, and whatever difficulty he or his students encounter does not make him any less enthusiastic about sharing ideas, playing with difficulty, and bringing out potentiality in each student.

As Petra Munro Hendry keenly observes, Doll "has an incredible ability to see the brilliance in someone, and to shape the diamond in the rough" (interview with Hendry, 2010). It is amazing pedagogical crafting to find the potential in a student and to bring it out before the student sees it. Many students need such nurturance before they can blossom. William Pinar also comments on Doll's devotion to students, "He was invested in what they were doing and what they were thinking, and he often emphasized their achievements and the progress they were making" (interview with Pinar, 2012). Pinar further explained that his devotion was so strong that it helped Doll to deal with the grief of divorce from Mary Aswell Doll during that difficult year, as students' achievements cheered him up.

"Passion for teaching often equates with the desire to make meaningful connections with others" (Mirochnik & Sherman, 2002, p. xi). Such meaningful connections with students to inspire them to move upward have been a key to Doll's pedagogy. As Kevin Cloninger and Christina Mengert (2010) point out, joy and creativity are linked in the human search for meaning beyond the self. The joy Doll takes in students' search for meaning helps them engage with him in a creative pedagogical process in which they are open-minded and willing to take intellectual risks. Students' curiosity, intuition, and emotional investment in learning can be activated by a teacher's joyful commitment to them.

As Hendry comments,

> I've never worked with someone or seen a fellow colleague who has the patience, loyalty, and commitment to the students that Bill Doll does. He shepherded through

several students who had taken the full seven or eight years to finish their doctorates and he never gave up on them. (Interview with Hendry, 2010)

Such a unwavering support and commitment to students comes from Doll's passion for life and his love for promoting students' growth. Not trying to control students' learning, he accompanies them in working through difficulty and takes time to work with them to reach another level of understanding and awareness. His sense of temporality is not linear but follows a process of emergence and self-organization. Getting doctoral students through the finish line at the last minute of seven or eight years (on one occasion, even longer) at LSU did not cause him stress but energized his persistent effort.

According to William Ayers (2002),

> Teaching at its best requires heart and mind, passion and intellect, insight and intuition, spirit, understanding, and judgment. Teaching, again at its best, can be an act of hope and love—love for persons, love for life, and hope for a world that could be, but is not yet. (pp. 39–40)

As a philosopher, Doll embodies his passion, hope, and love in his teaching to ignite students' passion for the yet-to-be. "A pedagogy that presupposes passion respects and makes space for our own and our students' intensity, energy, and honest engagement for the sake of understanding and meaning-making" (Romano, 2002, p. 376). Those who passed by Doll's classes would often be impressed by the shared laughter, energy, and intensity among the professor and students that overflowed into the hallway.

> When I was a doctoral student, Dr. Doll often cheered me up with his joyful demeanor. His joy in my intellectual capacity was always explicitly expressed, often unexpected by me. In traditional Chinese culture, one's parents and teachers seldom praise their own children, unless something exceptional happens. Not accustomed to such overflowing praise and affirmation, I was nevertheless much encouraged as I ventured into a new world. I shall never forget the big smile on his face when I told him that one of my papers had been accepted to be published as a journal article early in my doctoral program. That would be my first publication in English. He was so happy that he hugged me when hearing the news and hugged me again when he left his office. Actually, he was the one who had encouraged me to send off the paper in the first place—I had not thought of publishing since I was still a relatively new international student. Every time he read my writing and really liked it, he looked happier than I was! His joy was persuasive, lifting up my spirit even when I was having a bad time for one reason or another. Throughout my student years and then into my years of being a professor, he has continued to mentor me. His ability to see the potential in me before I had any idea has been amazing. Over all these years, for my every important step along the way, he has always maintained his enthusiasm and support for my work, including writing this book.

Doll's presence and leadership in the curriculum theory community at LSU has left a distinctive mark on shaping a culture that "really was around graduate students. It wasn't around the faculty. It was around seeing the graduate students as peers, as budding intellectuals who we could learn as much from as we could teach," as Hendry recalls (interview with Hendry, 2010). Curriculum theory professors facilitated and mentored students in various ways, one of which was an annual conference they organized for graduate students. The Curriculum Camp was first a local conference in Louisiana organized by LSU and the University of New Orleans and then became a regional or even national conference, including participants from other states. At the conference, a well-known scholar was invited to give a keynote address as the Campfire Speaker, but only graduate students were allowed to make presentations. It was intended to help graduate students see themselves as scholars in a safe environment where they could explore ideas without feeling intimidated.

William Doll initiated the Curriculum Camp. Doll had been its main organizer for many years and tirelessly brought in the younger generation to become familiar with the advances in the field through listening to the keynote address and to realize their own potential through peer interactions. I remember my first Curriculum Camp experience and how it felt to be part of an intellectual event, speaking at a conference as a first-year international student. For many students, attending the conference was not only an empowering experience of speaking in their own voices (Miller, 1990) but also a communal process in which inquiry was not (merely) solitary but entered into with companionship. Doll's joyful commitment to students went hand in hand with his passion for exploring ideas in a community.

Throughout his life, Doll organized one group of scholars and students after another to explore all kinds of interesting ideas, many times around dinner tables or at parties (see Chapter 7 for more details). Al Alcazar greatly appreciated a professor asking students to have lunch together and started to have dinner with his own students after he experienced the effect of such an arrangement. For Alcazar, Doll's sharing meals with students while engaging in intellectual conversations disrupted the hierarchical pedagogical structure and he found it empowering. Doll invited everybody to be part of a circle, invited everybody to become passionate about intellectual life, but he did not impose his ideas on them. It was always by invitation (and not a requirement), and Alcazar appreciated the invitation (interview with Alcazar, 2012).

Passionate Teaching

I am joyous, because every time I come into the classroom there is some problematic for us to wrestle with that came out of the students' concerns. (Doll, 2002a, p. 128; italics in the original)

According to Deborah L. Ball and David K. Cohen (1999), a recurrent problem in teacher education is that it does not provide an inspiring pedagogy that can counteract "the powerful socialization into teaching that occurs in teachers' own prior experience as students" (p. 5), but instead serves to reinforce traditional approaches to teaching. Pre-service and in-service teachers need to move beyond their own experiences of schools in order to explore new possibilities of teaching (Pratt, 2011). Under the current pressure for standardization in education, such uninspired pedagogy may only become intensified. But that did not appear in Doll's classes; Doll's students consistently remarked that his teaching was inspiring.

Such inspiration is intimately related to Doll's passionate teaching, which inspires students' passionate engagement with learning. Alcazar found Doll "always fascinated by exciting, new ideas, and that is what I love about him" (interview with Alcazar, 2012). When an instructor is passionate about what he teaches, his passion can influence students' relationship with what they learn. As Romano (2002) points out, "Passion sustained within [teacher educators] gives us pause to recognize the passion in our students, and in recognizing it, nurture it" (p. 374). Recognizing students' potentiality and creating the pedagogical conditions for students to realize it characterizes Doll's teaching as a teacher educator.

Doll considers passion at the heart of teaching and learning. In a doctoral students' orientation class at LSU, when it was his turn to talk with new doctoral students, he brought Donna Trueit's photo into the classroom. Pointing at the photo, he said: "This is my syllabus! It is about passion!" I cannot say I was not a bit shocked by his unconventional style, as it was my first semester studying in the United States, but his joy was contagious. Students were laughing together with him, and the class conversation was filled with vital energy. His ability to reach out to diverse students is partly linked with his optimistic and joyful spirit. Who can resist a teacher who always sees the good in you and joyfully tries his best to bring out the best in you?

The excitement that Doll brings into a class is his passion for students' generative knowing process, their personal growth, and a complex class conversation. He directs students' attention to their own intellectual interests and experiments with different ideas. He also understands that "the binding force of passion is counterbalanced with the liberating force of play. The tension between these two, an essential and productive tension, produces that 'third space' where newness, creativity, generativeness reside" (Doll, 2003/2012, p. 110). However passionate, the teacher cannot be carried away by passion, but must have a playful spirit to negotiate with boundaries, ideas, and relations so that students' own explorations and creativity can be inspired.

Passionate play with ideas in the classroom is intimately related to "crafting an experience":

> We [as teachers] can encourage students to "plunge into" subject matter, to see, feel, experience its aesthetic qualities—to explore the *spirit* of the subject at hand, as it were. And, in this process, as the experience begins to "overwhelm them," the students need to summon their own creative energy and thus help direct the experience to "its own end." (Doll, 2004/2012, p. 99; italics in the original)

Such a process of immersing oneself in a topic and responding creatively demonstrates students' intellectual presence in what they are studying and brings aesthetic and spirit-ful qualities to their learning. The interaction between passion and play is a whole-being experience, in which passion does not overwhelm but releases students' creative energy, and play does not go to all directions but improvises new responses to the issue at hand. The teaching episode in Chapter 1 demonstrates such a process in which students craft an experience by fully exploring a notion from different angles—hermeneutical, narrative, comparative, or philosophical—and direct such an experience to its own end.

A pedagogy of passion is an embodied way of teaching. Alcazar remembers how "the waves of laughter" filled the restaurant when Doll had lunches with students (interview with Alcazar, 2012). Sharing food and sharing ideas are inseparable for Doll, as both mark the enjoyment of life. It is also significant that he was able to bring these two together more fully in Louisiana where there is a culture of hospitality around food and gatherings. As William Pinar puts it, "It is odd that Bill saw Louisiana only in positive terms" (interview with Pinar, 2012). Susan Edgerton (1991) wrote about her complicated feelings about her home, Louisiana, with its beautiful countryside, sensuality, music, food, and Southern hospitality but at the same time layered by racism, sexism, and poverty. Not that Doll was unaware of the history of Louisiana, but he took great pleasure in the local culture, and its shadows did not prevent him from enjoying its positive aspects.

In the classroom, although Doll seldom organizes hands-on or visual activities that are the conventional modes of embodied teaching, he incorporates intellectual and spiritual play into the classroom, so that embodiment is introduced into teaching and learning at another level. In his classes, most of the time students converse with him and with one another, but in conversation, meaningful sense-making occurs, community building happens, relationships are sustained after the class ends, and the joy of stretching one's mind is shared. As Trueit (2002) points out, a conversation is "linked to balance, breathing, and the bilateral symmetry of the body" (p. 274). In Doll's classes, conversation is also linked with the dramatic performance of the body in laughter and the outpouring of positive energy.

Mary Aswell Doll (1995) comments that passion should be "blue fire" that ecologically performs the function of connecting, rather than "red fire" symbolizing repressed anger. "To be cut off from one's passion is to be cut off from one's emotions" (p. 127), so she argues for introducing blue fire into the classroom, to evoke and energize students' learning, as "laughing about serious matters is what makes us human" (p. 128). Such a laughing spirit makes it necessary to question the status quo and to renew lost relationships. Scholars (Britzman, 1991/2003; Chodakowski, Egan, Judson, & Stewart, 2010) argue that the field of teacher education neglects the role of emotions. Adequately addressing students' emotional well-being necessarily involves dealing with vulnerability and anger in addition to sharing joy and humor. Doll is sympathetic to and patient with students when they encounter difficulty, but he also has faith in students' ability to emerge from difficulty with more strength.

Doll's passion for a meaningful intellectual life is also reflected in his ability to listen to others deeply and ask interesting and evocative questions to keep conversation going. Mary Aswell Doll (1995) tells us: "If we are to experience the blue fire of the cosmos, we would do well to listen, pay attention, and recognize that the spoken language is only one of the speakings all around us" (p. 133). Silence also can speak. Whether at a conference, in conversation with a colleague, or in class, Doll is able to take in what others try to say or not say and engage them through questions or comments. Sometimes he closes his eyes while listening, but his ears are open and receptive, and when he opens his eyes, he always has something intriguing to say.

Trueit comments, "His ability to listen to the information and come up with a very cogent set of feedback, even though he looked like he was sleeping, is just amazing" (interview with Trueit, 2010). She also adds that this intellectual improvisation is something that he has always been good at. It draws students into the conversation and they come out of dialogical encounters with a more advanced level of thinking. It is through this passion that Doll managed to bring intellectual rigor to the coursework of Holmes's students at depths that were unusual.

While Doll is not much concerned with the question of subjectivity, students' interior worlds are touched by his teaching, so a space for subjective and intersubjective transformation is opened up in class. As Trueit points out, "Bill's notion of the self is the self-in-the-system" (personal communication, 2010), so one works on oneself through working with the system to create a space for personal growth and transformation. This sense of the self-in-the-system is consistent with the ecology of selfhood in Bateson's view and the fractal self in complexity theory, neither supporting the notion of a separate entity. Thus, Doll's passionate teaching sparks and informs students' learning through the network effect but does not center on the teaching self.

Passionate Learning From China

> [T]he binding force of passion is counterbalanced with the liberating force of play: The tension between these two, an essential and productive tension, produces that "third space" where newness, creativity, generativeness reside. (Doll, 2003/2012, p. 110)

Learning from others has been a consistent thread through William Doll's life history as a student, a teacher, a principal, and a teacher educator. Part of the joy in teaching, for Doll, has been that he can constantly learn from students and colleagues. Becoming a student of students is an important aspect of the teacher's—and the teacher educator's—journey (Ayers, 2002). Doll also has a passion for team teaching. He commented on his co-teaching with Hendry: "That was a total and complete joy! It was just fun!" (interview with Doll, 2010). In the Holmes program he put students in pairs to work at school, and he loves to pair with another instructor to teach.

Sharon Todd (2003) makes a distinction between learning *about* the other and learning *from* the other: The former runs the risk of objectifying the other while the latter indicates the necessity for mutual engagement. On most occasions, I think, learning from the other already incorporates learning about the other because knowing the other is always a component, but we must go further to let the other's teaching change ourselves. Doll's passion for learning *from* the other is more dramatically demonstrated in his encounter with China. After his first trip in the fall semester of 2000, he continued his engagement with China for more than a decade, an example of his passionate learning from the other.

Encountering the Other

As his graduate assistant, I accompanied Doll and Trueit to China in 2000 and assisted with translation and lecture tour arrangements. We stayed for two months. The timing of his visit was at the initial point of the United States movement toward the internationalization of curriculum studies. In April 2000, Pinar and Doll organized an international conference on the Internationalization of Curriculum Studies at LSU. It was decided at the conference that the International Association for the Advancement of Curriculum Studies (IAACS) would be formed and that the location of the next IAACS conference would be in Shanghai, China, in 2003. The purpose of this international association is to promote scholarly conversations across national borders while counteracting both nationalism and globalization, marked by uniformity, in order to create a worldwide curriculum studies field (Pinar, 2003).

It was in this context of excitement about the international landscape that Doll embarked on his journey to China. I still vividly remember how excited he was by the big adventure as we were flying into the Pudong International Airport in Shanghai. He was immediately attracted to Shanghai—our first city on the tour—and to what was happening in China, even though it is a very different country with a very different culture from that of the United States. I was amazed that he had such an instant connection with Shanghai. Many aspects of the two countries and two cultures are opposite to each other (Wang, 2004, 2014) but apparently the differences were positive to Doll.

Upon his arrival, he was curious about everything, attending to the smallest details that appeared different. He stopped on the street to look up at the tall apartment buildings where clothes were hung outside the windows to dry. He observed that traffic did not appear to have clear patterns by American standards. He and Trueit took the subway without me to try to navigate on their own. In the subway and on the street, Doll could almost always find help when needed because many Chinese, especially young people, could speak English. Recognizing that the Chinese culture has a strong tradition of hospitality to guests, he still felt particularly honored that ordinary people would go out of their way to help him.

As he visited seven different places in China, from the north to the south, and experienced different landscapes, lifestyles, and cultures within the same country, he was quite taken by the dynamics and the vibrancy of this nation. In particular, he realized that China, with a history and mentality so different from that of the West, has a lot to offer to the world. In this sense, being open to "the otherness of the other"—the Levinasian and Derridian notion—was not an abstract idea but became concretely embodied in his understanding of and learning from the Chinese culture and people.

At that time Doll was reading Daniel Bells's theory about postindustrial society. As the plane was landing at the Pudong International Airport, he looked at the agricultural fields from a distance and then looked at the airport—its contemporary architecture showcasing Shanghai's cosmopolitanism—and it hit him that here was a country combining the preindustrial, the industrial, and the postindustrial all at the same time. Doll's (1993) first book discussed three stages of intellectual and cultural development in the West: premodern, modern, and postmodern; yet it was clear to him that in China all three coexisted. He pointed out in one of his letters from China to friends, relatives, and colleagues,[1] "The contrasts between the old and new China, age and youth, is most dramatic; more so than in any other country I have visited."

Doll enjoyed various examples of postmodern architecture in Shanghai—he already knew about them before coming to China—and also saw how they juxtaposed the standardized routine of schooling and workers' and farmers' daily struggles. He described Guilin, a southern city well-known for its landscape and crystal clear water in the Li River, in a letter: "Bamboo growing, farmers using oxen for plowing the fields, men using cormorants for fishing, and women washing clothes by hand along the river banks." He also saw families living on boats. This closeness to the earth was hardly as romantic as portrayed in poetry and songs; it was part of a harsh life for many people. Such a paradoxical coexistence of cosmopolitan postmodernism and a premodern lifestyle intrigued rather than puzzled him, and he exclaimed in his letter: "Here is a China not emerging but racing from the past into the future!"

With a cosmopolitan orientation (Hansen, 2011), Doll has a passion for the best humanity as it is reflected in different cultures at the crossroads. His version of cosmopolitanism recognizes that the positive role of difference and "ethical engagement with difference" (Pinar, 2015, p. 36) in the internationalization of curriculum studies requires dynamic interplay so that difference is neither excluded nor self-contained. With Doll's interest in chaos and complexity theory, he was fascinated—rather than disoriented—by the scenes unusual to Westerners' eyes. Where others might see chaos, he saw self-organization at work. He was also impressed by the role of negotiation in China: Chinese negotiate everything to bend the rules. After the initial surprise, he could appreciate the beauty of such disorderly order.

In one of his letters back home, Doll further commented:

> Often I feel like Alice, in a strange and intriguing land, not quite understanding either their traditions or my own. All is topsy-turvy, as seen through the eyes of my Western logic—and indeed I am seeing yet again, that there are many other ways of reasoning besides the Western. (Letter from China, Doll, 2000)

This sense of being open and learning from other ways of reasoning is related to his ability to allow perturbation to play a positive role, his full presence to what is different, his willingness to play with the unfamiliar, and his insights into patterns that are nonlinear and irregular. In this sense, his pedagogy of passion is connected with his pedagogy of play, perturbation, presence, and patterns. His encounter with China unfolds a process of self-education through his passionate learning from the other. Such self-education means not only understanding the other but also understanding the self in a new light. This gesture of "turning inward" to question the self is an indicator of mutuality in his engagement with the other.

William Doll saw the Chinese as hard-working and uncomplaining people who, in turn, questioned his own limits by their very existence. In his own words,

> I remember it in 2000, people got up early and worked at their tasks. I remember going along one of the side streets by the hotel and the little shopkeepers would have a bed behind it and that's where they would spend the night and then they would come out and do their shopkeeping. I remember people with old, ancient bikes piled high—one of them was a mattress. So I found a hard-working people who seemed to appreciate life and living and were not asking for privilege. (Interview with Doll, 2010)

During his stay in China, he questioned his own privilege. In the letters he wrote back home, he kept asking why he should be the one who needed extra care in the hotel. Of course the extra care in China was perceived as basic living conditions in the United States, such as hot water, heat in homes, or toilet paper in the bathroom. The use of electricity for heating during the winter was monitored by the Chinese government to save energy. Although acknowledging that the trip might have been a superficial experience, as he was at the receiving end of a warm reception, the contrast was dramatic. Witnessing Chinese people's way of life, questioning his own privilege, he asked a fundamental question: What does "civilized" mean?

As Al Alcazar points out, encountering and learning from the other is a journey to wholeness of spirit (interview with Alcazar, 2012), and Doll's journey to China has been such a process. The wholeness of spirit is Alcazar's attempt to blend the Deweyan spirit and the native notion of the communion between human and nature in the Philippines with a subtle critique of Western human-centeredness. Wholeness is not within the self but in relationships. Seeking spiritual wholeness, Doll engaged in a process of expansive growth in the Deweyan sense (Garrison, 1997). In this journey, he accepted the challenges that Chinese life posed to Western thinking and, after his encounter with the other, started to rethink the meaning of a curricular sense of story, culture, and different ways of seeing.

The Interplay Between the Self and the Other

A dynamic interplay between the self and the other not only requires a reexamination of both the self and the other but also invites the other to reexamine its own self so that both the self and the other benefit from the process. The *mutuality* of cross-cultural, intercultural, and international engagement and conversations is the key for enabling transformative change of both the network and its participants (Wang, 2014).

During his trip to China, Doll made the connection between ancient Chinese garden art and the postmodern Western fractal, and marveled at it in one of his letters: "Here the ancient Chinese culture, most intricate and routinized, is also beautifully irregular in its order, shown artistically in the landscaping, rock formations, gardens, and serpentine pathways." While acknowledging the hierarchical nature of the Chinese political system, he was intrigued by the complexity and naturalistic principles of Chinese aesthetics. In making such connections, he offered a stranger's eyes for the Chinese to see their own culture as if for the first time.

As I accompanied Doll and Trueit in touring Chinese gardens in Suzhou and Hangzhou, he commented throughout our time together about how marvelously compatible these gardens' designs were with the principles of fractals. Later I wrote a chapter in his co-edited book to discuss such a link in the context of curriculum theorizing (Wang, 2005). He saw fractals everywhere in those gardens and was amazed by how traditional Chinese aesthetics demonstrate the wisdom of fractals, which have been only a recent area of inquiry in the West. Doll was aware of Chinese philosophy and aesthetics and Western chaos and complexity theory connections that have been made in the literature (Briggs, 1992; McGuire, 1991; Walter, 1994), but seeing the connections with his own eyes was revelatory for him. His keen observations gave me a fresh lens through which to reexamine my own culture.

Coming back from China in 2000, Doll felt that he had a deeper understanding of the Derridian notion of respecting the otherness of the other (Derrida, 1992), as his experiences in China enriched and embodied such an openness to the other. He also urged the Chinese to learn from the best of their own traditions. As Doll traveled to China over the years, he also worried that the Chinese would take on the worst of the West in neglect of their own strength as a people. In her first encounter with Doll at an international conference in 2003, Jie Yu was impressed by his deep respect for another culture. As he visited different museums in China, he hoped that

> the Chinese culture could be lived in the daily contemporary lives of the Chinese people so that it could be seen in the school background, in parks, and on the street, rather than merely in those historic museums. (Personal communication with Yu, September 29, 2010)

To regenerate the best part of a heritage in a global setting is a challenge for every nation, but particularly for China with its complicated and difficult historical relationships with the West since the mid-nineteenth century. It has always been my own vision that if the best of the West and the best of China could be integrated, all would benefit tremendously. It is a question that has been asked in China in various ways and various answers have been attempted (Wang, 2014). Perhaps now it is also the time for the West, for the United States, to ask a similar question: How can we open doors to learn from the other and integrate the best from all?

Doll's passion for China was reciprocated by the Chinese, because Chinese students and professors alike were amazed by his outpouring of enthusiasm and his appreciation of Chinese culture. They had already studied his postmodern theories both in education and in philosophy. Doll was warmly received intellectually and culturally. He was taken by how much the Chinese were interested in his work and their willingness to learn from his theories to inform the Chinese situation. The recent Chinese curriculum reform has attempted to break through the traditional education that centers on exams and to promote individuality, creativity, and flexibility (Pinar, 2014). Doll's theory, in combination with his vibrant personality, breathed new life into the climate of reform.

The warm reception in China was a pleasant surprise to Doll because, despite his achievements, he had felt like an "outlier" in the U.S. field because he did not fit into any one school of thought. Making a contribution to the Chinese field in ways that were beyond his expectations stirred his enthusiasm. Since their first academic visit in 2000, Doll and Trueit have returned to China a number of times. In 2014, Doll was invited to visit a Chinese elementary school to help teachers implement the 4 R's in their teaching.

Over the years, Doll's understanding of Chinese culture and Chinese education has deepened and his passion for learning from China—ever changing at a fast pace in the twenty-first century—has not faded. The mutual enthusiasm between him and Chinese educators has created a dynamic interplay that promotes wholeness on both sides in an ongoing process of engaging the worldiness of curriculum studies (Miller, 2005; Pinar, 2009).

Passionate Teaching in the International Context

Doll has taught in several countries outside of the United States. He had taught regularly in Canada during the summer while he was a professor at LSU. He visited Russia and Finland for academic purposes. Now he is teaching at the University of British Columbia. Although the occasions for international teaching have varied, he has never significantly changed his teaching orientation. In Canada, expectations from the university put some constraints on his way of teaching, but in terms of cultural differences, teaching in Canada cannot be compared with teaching in China. China is more dramatically different. However, Doll's nonlinear teaching style and his passion for teaching traveled with him to China.

When Doll taught a one-month graduate seminar in Shanghai at East China Normal University in 2000, he did not change the fundamental elements of his teaching, even though Chinese students had been accustomed to teacher-centered methods. Their expectations from a professor and their capacity for engaging in academic conversations were different from those of American students. Doll creatively facilitated the discussions; he did not lecture to cater to what Chinese students were familiar with but rather ran the class in a way similar to how he taught American students. Nonlinear by nature, his teaching still reached Chinese students, as they learned to express their own perspectives.

I think that for students from diverse backgrounds, the teacher's passion for teaching and genuine concern with students' ideas, well-being, and progress can transcend cultural boundaries to inspire students' engagement with learning. Doll's different teaching style became a positive factor because Chinese students arrived in the classroom expecting a different approach to teaching from an American professor. They came with an open-minded attitude for the unexpected, so they had a high level of tolerance for what was difficult for them. They were eager to learn from the West, particularly from the United States.

Echoing American students' appreciation of Doll's teaching, Chinese students particularly appreciated Doll's enthusiasm for students and positive affirmation of them, his passion for exploring ideas, his ability to build connections between different perspectives, and his unorthodox and humorous teaching style. To a certain extent, they found his teaching liberating from traditional constraints of Chinese teaching, as it was so different from anything they knew. During that one-month seminar, I observed the class dynamics and was impressed by the Chinese students' creative energy demonstrated under Doll's guidance. Even though language posed a challenge to some degree, ideas flowed in class interactions.

Since Doll's first visit to China, visiting scholars and graduate students have come to the United States and Canada to study with him. For example, in the class he co-taught with Trueit at the University of British Columbia in 2013, almost half the seminar participants were from China. Doll has long been interested in understanding why Dewey had been so excited by his own visit to China from 1919 to 1921. Dewey had planned to visit for two months but ended up staying for two years. Doll wondered whether Dewey's experience in China had any impact on his philosophical thought. Although many studies explore Dewey's influence on China, few consider China's influence on Dewey. Doll believes that the visit must have had some influence on Dewey's later thought. One of the Chinese students in that class at UBC, Guopeng Fu, explored Dewey's experiences in China. Following is a teaching transcription that shows how Doll led a class discussion on Fu's paper.

Fu: Originally I wanted to write about the influence of Dewey on China. Then, with Donna's suggestion, I changed to what events influenced Dewey. I still find it very difficult to do [*laughter from the class*]. I am studying it from four aspects. One was his frustration that World War One did not end as he wished before he went to the Far East. I would say it was a part of his motive to go to Japan and China. The second was his encounter in Japan, which was related to what he would face later in China, as the students' movement was a result of the conflict between China and Japan. The third aspect was what happened in China while he was there, sort of the rebirth of a new nation from his perspective. And the warm response he received from Chinese students and intellectuals, because he was perceived as a figure of Mr. Democracy and liberal....

Doll: That is a nice phrase! Mr. Democracy!

Fu: So that also excited him as a scholar. He was warmly welcomed into China and he witnessed the shift in China initiated by students. The last aspect was his encounter with Chinese history and traditional thought such as Taoism and Confucianism, which sort of informed his later intellectual development on natural piety. These are aspects that I would like to look at, his experiences in China rather than the influence of China on him. I don't want to go through all the literature and write a book like that [*pointing to a big book*] [*laughter from the class*].

Doll: Are you going to read it?

Fu: Oh, yeah, yeah, I will read it.

Doll: In that book, there are a lot of papers on China. You can pay attention to it. You also can look at letters. I may be wrong, but not many people pay attention to them, at least not that I have come across.

Fu: The letters are like a three-hundred-page book. The first half is about his experiences in Japan, and the second part is about his China experiences, but it was only three months of letters, even though he stayed in China for two years. I missed the other part.

Doll: But I don't think people have paid enough attention to it. That is my question.

Fu: I am not really sure, but yeah, that is good to know.

Doll: Let's assume that it is true, and then it is one of the interesting things about your paper. The four aspects you have chosen about his visit in Japan, his displeasure, and his reactions, and then his visit in China and his reactions. And then you play off the two. Finally the last one is the relationship between Dewey's thought and the history of Chinese philosophy. I think that is very good! In terms of your writing, among other things, you came up with lovely phrases.

I think you need to tell your stories with a bit more of story-ness, if I may say. In the first paragraph there are a whole bunch of facts. It would be nice if you have in that paragraph, actually, a couple of pages

	in which you tell his story about his unhappiness. You have facts down but not stories. I am trying to move you, if I can, from a sense of facts to a sense of story. Because stories make things come alive. You can make it a little deeper, a little more intriguing, you add enough color to it so that it begins to captivate your readers. Putting out of facts does not captivate readers. Telling stories does.
Fu:	My concern is that I don't read that [type of writing] really often in academic....
Rasunah:	But you know the story. The story is that he went to Japan, and then to China, and he learned all these things, right? Then he started to synthesize and respond to his experiences, right? So your stories are already there.
Fu:	Okay, if you think about it like that....
Rasunah:	This is about a human being....
Jung-Hoon:	Are those two elements in chronological order? If so, then you can make it flow.
Fu:	Well, I got the idea from one of my readings in which an author commented on China, but not necessarily presented it in a chronological way.
Doll:	But the events are chronological. In your last event when he encountered Chinese philosophy, certainly it did not occur after he visited China but occurred while he was in China. You move to the May Fourth movement too quickly. You need to story his unhappiness—you merely talked about it. In this first paragraph, you set things up, which is nice. Set it up a bit more slowly. It is a lovely beginning, but it is too factual. A story moves beyond fact; it uses facts and incorporates facts, but there is a flow to it. The other day when you asked me how I moved from pragmatism to postmodernism to complexity theory, and you finally decided that it was just flow. Let the paper flow. That is a different type of writing, but it is a writing that is, I think, going to appeal to your readers. You need to make facts come alive. That is the art of a story.
Rasunah:	Yes, we need those stories.
Fu:	I'd really like to achieve such in my dissertation. I was reading those interviews with teachers and felt inspired by their stories. But I find it really difficult to put those stories into words, and expectations for analysis are different. When does it become analysis? I want to find a way to demonstrate teachers' stories in my dissertation. When you read a dissertation, however, it is more about information and knowledge.
Doll:	You just need to spend more time with Carl Leggo! [*laughter from the class*] He does stories. For the reasons that you mentioned, it is difficult to write such a kind of dissertation. You are in a place where going beyond a dissertation is allowed and where people do different things.

	Now try it. Of course you don't want the stories to get . . . well, a story lives a life of its own, and you can become part of it and you work along with it. You cooperate with it. You have to make sure that it is not too much of you looking at yourself and your writing. I don't think that happens here but that can happen . . . . Bringing forth narratives and you have people here doing narrative work.
Brenda:	I am wondering, how many years have you stayed in Canada? Two years?
Fu:	Five years.
Brenda:	Okay. Dewey stayed in China for two years. I wonder if you can bring your stories in: What it was like for you to come here from China, and intermingle them with Dewey's stories. I think that would be interesting for readers; it would be very personal, and it would be autobiographical as well as biographical. I found out on the Internet that lots of people have written about Dewey in China, so what makes your stories unique?
Pete:	How about you write some kind of letters to China? [*laughter from the class*]
Brenda:	And they don't have to be real.
Doll:	To put some limit to these ideas . . . . These are fascinating ideas! I would take a guess for a journal article you would stay within a certain word limit. For some other times, you may use and develop these ideas. And they may appear in your dissertation, maybe not exactly as suggested, but I think they could have some kind of effect. Particularly you want to talk with teachers. Here you might be able to insert yourself and your experiences in a very meaningful manner. For this paper, I would not suggest it. It could show up elsewhere.
Fu:	By the way, I want to respond to Brenda's idea. There is a visiting scholar from China who is studying Chinese students' experiences here. I just got interviewed by her. She asked a lot of questions that I never thought about—they are not stories in her writing but the interview made me think about what I went through and what kind of transformation happened to me. That was really interesting. I never thought of things that way.
Doll:	Hongyu does this beautifully. I don't have the title exactly, I apologize, Hongyu.[2] She talks about coming home, the plane landing, which was the beginning of exploration of the sense of the self through Kristeva and others, but she does that with engaging stories. That title is evocative: Who is the stranger? What is the call?
Fu:	It also happened to me. When I returned to China, my buddies asked me all kinds of questions: What is the difference after my studies here? The notion of the stranger is interesting.
Doll:	Let's go back to Dewey. He visited a lot of countries, but China was different. He did not intend to spend two years but ended up spending

	two years. The whole journey was a longer journey and there were reasons for that. You begin with a sense of disappointment, then let this disappointment come forward. Not merely factual, but you need to let us *feel* his disappointment, and his hope, and a world democracy was his hope. You have good authors to reference with. And why did he pick Japan to go to?
Fu:	Oh, he . . . well, it is a rhetorical question. I don't need to answer it [*laughter from the class*].
Doll:	Yes. You also need to help us understand why he went to Japan. That was clearly a mistake for Dewey in all kinds of ways. We need to feel why he chose Japan, and why he was unhappy there, and revelations that happened to him. If he had not gone to Japan, he might not have become enthusiastic about China. And if he had not arrived in China a few days before the May Fourth movement, which goes back to your statement, then there might not be a sense of bringing back those hopes he had. It is going to be that sort of weaving, to weave these stories together. There is a sense of chronology but there is also a sense of recursion; like in any good rug, we see recursion with something new. Particularly an interesting rug. So make it interesting.
Fu:	Sure, I will do my best.
Doll:	Whoosh! A lot of good stuff! (Video transcription, March 23, 2013)

Here we can see that Doll's passion for the topic was woven with his own scholarly inquiry. He posed detailed questions to Fu and guided students' inquiry. Conversations among students were often humorous, filled with laughter, serious probing, and imaginative ideas. I have to admit that I was a bit surprised by Doll's push for Fu to engage in storytelling in that particular paper, because although narrative inquiry was in general accepted by UBC professors, the broader academic field might or might not respond warmly to such storytelling. In my own experience, I seldom witnessed Doll asking a student to follow a particular direction in his or her work so specifically. Certainly his passion for the topic was tangible and his enthusiasm for story as a way of writing and presentation—one of his 3 S's—was persuasive.

Conversations among students in the class were particularly interesting as students helped to show Fu how to narrate experiences and suggested imaginative ideas to creatively deal with the issues at hand. They also suggested that Fu make a connection between his own experiences of studying in a foreign country and Dewey's experiences in China. Although the imbalanced relationship between China and the West makes such a comparison complicated, these ideas helped Fu situate himself in his transnational and intercultural context. The relationship between narration and analysis in writing was another generative topic. Students' intellectual open-mindedness and playfulness with each other would not have emerged without a sense of community. In this sense, passion was shared in the class.

Just as Brenda asked whether Fu's own experiences could be informative, I wonder whether part of Doll's passion for understanding the influence of China on Dewey is related to his own autobiographical experience. After all, the dramatic change occurring in China when Doll started to visit China has some parallels to what Dewey experienced there from 1919 to 1921 when the new culture movement was under way (Wang, 2014). It was the dynamics in China that captured Doll's attention, however, not necessarily world democracy, because Western democracy has been seriously questioned since the post-structural turn of events. Moreover, Doll's warm reception from Chinese scholars and educators and their welcoming of his curriculum ideas in Chinese curriculum reform echoed John Dewey's warm reception in China when democracy and science were perceived by progressive Chinese intellectuals as important for Chinese reform at that time.

However, Doll's concern with the *mutuality* of intellectual and cultural exchange is most important to notice, because the lack of attention to China's influence on Dewey's thought implies a one-way relationship in which the Chinese learned from Dewey but Dewey did not learn from them. Doll does explicitly acknowledge, both in writing and in teaching, the influence of his experiences in China on him as an educator and as a scholar. A pedagogy of passion through ethical engagement with difference requires critical self-reflection.

NOTES

1. When William Doll visited China in 2000, he regularly wrote letters to friends, relatives, and colleagues in the United States to share his thoughts about and responses to what he saw and experienced. Donna Trueit helped me locate four of his letters, although he had written more. The quotes from these letters are not dated because dates were unclear.
2. I requested the videotaping of this class when Doll and Trueit were teaching at the UBC. Once in a while, Doll would talk to me through the video—even though he could not see me and I could not respond—as if I were present in his class. It was funny for students but it also demonstrated Doll's awareness of the other even though I was absent in person. This is also an aspect of his presence to others.

· 6 ·

PEDAGOGY OF PEACE

Peace is usually not associated with William Doll's teaching, yet a sense of peace with himself and with what is happening in the world around him marks his pedagogy. I intentionally place this chapter after the discussions of his pedagogy of passion, because the interplay between passion and peace creates a unique pedagogical effect. As Jane Addams (1906/2007) notes, new ideals of peace are not tranquil or passive but complex and dynamic, leading to the nurturance of public life. Doll's pedagogy of peace is dynamic and closely related to his sense of play (see Chapter 1). Playing with the limits and playing with social relationships would not be possible without a certain flow that springs from inner peace.

Doll does not like to talk about his difficult teenage years. But those years were foundational, because he made peace with his own eccentric styles. He reflects,

> I guess I spent much of my youth trying to prove who I was, partly to me, and partly to others, but I really felt tremendously inferior in school, and outside of the school I felt a very strong sense of self and in what I could do. (Interview with Doll, 2009)

During his teenage years he found his niche outside of school. He retreated from schooling and created his own space. Those years were also marked by trauma and loss that he had to deal with (mostly in silence) in finding ways to become his own person.

The freedom to explore as a child and his tumultuous teenage years deeply influenced Doll. By making peace with his own losses and differences he became a dynamic yet calm person who could negotiate conflicts, turbulence, or disputes for positive change. As an educator at peace, he has helped his students deal with difficulties. Such a pedagogy of peace does not teach students directly; instead, students experience peace through their relationships with the teacher to enable learning.

Making Peace With Loss, Guilt, and Difference in Life

> We have lost our sense of certainty, but in this loss we may well find a sense of ourselves. (Doll, 1990/2012, p. 152)

Loss and Peace

Doll suffered traumas as he grew up. The loss of his father when he was twelve years old was a great trauma, especially because he remembers his father as a benevolent and gentle person who loved him and allowed him freedom to explore. One cannot help but wonder if it is a romanticized memory; after all, Doll did not know him well. That relationship was cut short even earlier, because his father became ill when Doll was only nine years old. His mother tried to find paternal figures for him as he grew up; he also found father figures in Colonel Kennedy, who employed him, and Steve Mann, who was his mentor. His father and these two figures were dramatically different, but he was able to incorporate all these differences and emerge as his own person.

Doll's father passed away under the shadow of World War II:

> Off the coast of New Jersey in the early part of the war, a German submarine surfaced. It didn't do anything but it was spotted. Hence, the whole East Coast became paranoid. We had blackout shades. If the whistle blew, we were to turn out all the lights in our house. When Father was literally dying upstairs, the doctor was there and his friends were there. They had the blackout shades down, but there was a little crack. If you had the light on, you could see a little crack of light. Whoever was the warden of the street came by and demanded that that light be turned off. Mother was furious, saying that there was a man dying upstairs. The warden kept saying that there was light. Father called down to Mother: "It is all right, it is all right, dear. We can turn off the light." Father died in darkness. (Interview with Doll, 2009)

No doubt it was a dark time for young Doll. When his father had his first heart attack, Doll was not aware of it because his parents kept it from him to spare him from pain. He was not allowed to see his father die but did attend his funeral. He was immediately sent away to a camp with his grief, loss, and anger not addressed. The camp counselor sent him back home because he was a troublemaker there. Addressing grief and loss both in family and at school is a difficult but necessary task (Burns, 2010). He and his mother seldom talked about his father's death, occasionally not talking at all, as his mother suffered from a huge loss, crying all the time. His father had waited a long time to marry his mother, and his parents were very much in love.

Another big loss[1] was the loss of Doll's teenage girlfriend. Doll was not a popular student, but found salvation in his girlfriend. However, after they had dated for more than a decade, she broke up with him because he did not want to marry until he was established. She soon married someone else. The loss of his girlfriend was so devastating that he felt stomach-ache. He sought the counsel of a priest, but the priest offered no help or comfort, simply saying: "Find a Catholic girl for yourself." He grieved the loss in silence.

Silence marked these losses. Doll did not receive the help he needed to address his pain. Dealing with loss silently perhaps gave him confidence that one can eventually emerge from the pain. Now he has faith in his students who struggle on their own journeys. However, not paying adequate attention to pain and its impact might have had side effects for Doll. Silence over loss does not help to build the bridge between affects and words for generating meanings (Kristeva, 2002), so difficulty quietly accumulates inside. Sometimes difficulty is expressed by acting out; Doll's behavior at camp is an example. I also wonder whether Doll's tendency not to dwell on negative aspects of reality is related to this pattern of silence over loss when he was young.

Trueit talks about how Doll tends to ignore his physical pain. She points out that he is in pain in order to draw his attention to it and then take the initiative to do something about it. It is not that he does not feel the pain, but that he does not seek the treatment that would alleviate it. He feels that he should be able to deal with it himself. For instance, for two years he did not see a doctor about a broken wrist that did not set properly on its own and later required surgery. She further points out that his attitude was influenced by the Puritan tradition of "duration and endurance" (interview with Trueit, 2010). Trips they had made together to visit Doll's hometown made it clear to Trueit that a particular cultural tradition had strongly influenced him, and it was interesting to her that he did not recognize such an influence. She points out:

> The hardiness came from the colonization of America, so this sense of strength and denial of pain are strong. Bill did not accept freezing for dental work until the last couple of years. . . . [When I notice he is in pain,] I have to push. He does not take initiative to ease the pain. (Interview with Trueit, 2010)

Doll's particular way of making peace with pain and loss was marked by both individual experiences and the cultural influence of New England. To cope with pain, Doll typically found external outlets to release internal energies. He worked in different jobs as a teenager and found meaning and fulfillment in the work. He dealt with difficulty by doing things. Even though this way of coping may have left him with a tendency to highlight the positive aspects of reality, it has also left a tender spot (not fully healed) in him that makes him sensitive and easily touched. As Trueit observes, Doll often becomes tearful when he hears stories about death, love, and loss (interview with Trueit, 2010). Therefore, his way of making peace with loss and pain has shaped a particular outlook that focuses on moving forward by engaging in meaningful activities.

Guilt and Peace

The teenage years are difficult for both teenagers and their parents, especially in American culture in which teenagers are expected to transition to being fully independent persons by marking themselves as different from others. Not popular at school, Doll had to deal with a sense of embarrassment when judged by others. His memory, however, is more related to feelings of guilt. He acknowledges his guilt for not being with his mother when she passed away in the hospital:

> The day before she died she said: "I want to go home; I want to go home." I told her: "Oh, you are fine and I will see you in the morning." Of course when I came to see her in the morning she was already dead. I never realized what she was asking for was to come home to die. That has stayed with me and will always stay with me. (Interview with Doll, 2010)

He wishes he had been more aware of others' needs so that he could have been there for his mother when she died. He felt guilty for being blind to her wish, and being a loyal person, failing his mother at the last moment has stayed with him heavily. Coming to terms with his guilt, Doll realizes that his sensitivity to others could have been heightened. He credits Trueit, with her amazing perceptions and phenomenal insights, for helping him become more aware of the other and how the other might perceive the world. With her help, he also realizes that his sense of guilt is related to his Puritan New England upbringing.

In much the same way that he deals with loss, he tries to deal with guilt by doing more to help others, but there is a downside, as both Mary Aswell Doll and Donna Trueit have told him: "Don't play God." He recognizes that in his efforts to help, sometimes he drives others away because he is doing too much. In pedagogical situations, when a student is not successful, it weighs heavily on him, and he keeps wondering what he could have done differently to help the student. He has a "paternalistic" tendency to help students that not everybody appreciates.

> *Interestingly, my own sense of shame shaped by Chinese culture and gendered context largely dissolved in Dr. Doll's presence when I was a student. Shame attacks a person's fundamental sense of themselves with a constraining power. The way I was brought up in China constructed an important layer of my identity in a strict sense of morality and ethics. Encountering crisis and difficulty in my cross-cultural journey, I had my moments of embarrassment and shame. However, I had not felt any acute sense of shame in front of Dr. Doll as I would in front of others, even if I did absurd things. While the coexistence of presence and absence, as I discuss in Chapter 3, provided a certain protection of my interior world, Dr. Doll's nonjudgmental attitude—a person can be nonjudgmental to others only if he or she has achieved a certain sense of inner peace—and his trust in others' goodness provided a solid foundation for my belief in his acceptance of me no matter what happened. Such a pedagogical effect of a teacher on a student is quite an achievement.*

However, humor and playing with irony sometimes work well in pedagogical situations. To make peace with his own racial, gendered, and classed privileges, Doll presents himself as a "dead White male," a line of self-mocking that he often uses to acknowledge his blindness and limitations and his guilt about his privileges that others do not have. He does not feel offended if others also use that term in a playful way. In this sense, he is at peace with himself as a human being with blind spots. Pedagogically, Mei Wu Hoyt, from a student's viewpoint, also took such joking as indicating a change in the teacher–student relationship from the hierarchical to the egalitarian. She comments,

> It is okay to talk about issues that may challenge the professor in his classroom, and I had nothing to hide or worry about. . . . I think that helped me to start open and honest conversations about difficult issues in curriculum and in schools. (Personal communication with Hoyt, 2010)

So Doll's use of an ironic description of himself can help students free themselves from the hierarchical pedagogical relationship to engage in conversations in a more open-minded way.

Difference and Peace

Peace is not the absence of conflict, tension, or difference (Addams, 1906/2007; Lin, Brantmeier, & Bruhn, 2008; Quinn, 2014). In fact, an active form of peace becomes possible only through working with discomfort, difficulty, and difference. Recalling his teenage years, Doll commented:

> I think I was a relatively happy and well-adjusted child going into the seventh grade. In the seventh grade, those traits of boyhood identity did not surface strongly in me. I never liked the scatological jokes. I did not like the sexual references to girls. I did not really have much upper body strength. . . . Mother bought me punching bags, trying to get me to do those sorts of stuff with them. It meant that I was never "one of the boys."

> I was not an athletic. I was tall; I was gangly. . . . So, for my sports, I picked up things away from the crowd, away from the jocks. I played tennis at the New Ocean House. . . . I had a little boat. I would row the boat all over the place. And I would swim. I was even asked to go out for the swim team at one of the country clubs because I swam fairly well. . . . What identity required was really being "apart from."

> I remember in my junior year of high school, when I was sixteen, a new girl came into class, one of the pretty people of that time. I invited her to a school dance. I was tall and handsome, so she said, "Sure." In two weeks, she called me and was very straightforward: "I have been talking with some other boys, and they said that you were not the person I should go out with." That hurt. (Interview with Doll, 2009)

Not being one of the boys, feeling hurt by rejection, Doll nevertheless found a group of friends on the street. He ran a hotdog stand with one of the street kids who had street savvy and came from an underprivileged background. Although their backgrounds were very different, they enjoyed working together. Starting at a young age, difference—and the tensions accompanying it—became a site of learning for Doll and such experiences were beneficial in his later teaching career. Retreating to working outside of school to find fulfillment, Doll developed a sense of self in delivering papers, mowing lawns, working at a pipe factory and a milk factory, peeling potatoes, working as a busboy in restaurants, and working at hotels. In so doing, he developed his identity outside of school:

> My sense of the self came not from peers, but from my working experiences in which I learned I could be somebody and I felt I could do things, and I could accept responsibility and carry through with what was being done. . . . I also learned a lot from different people . . . and learned to work with somebody else. (Interview with Doll, 2010)

Doll started to work when he was only eight years old; in the years following, among his many jobs was that of car salesperson. Being a salesperson means often dealing with a sense of failure. His early encounters with an ongoing process of letting go of failure to try to achieve the next goal proved invaluable in developing his skills in ongoing engagement. While working for Colonel Kennedy, who ran a hotel business, Doll observed the art of negotiation:

> When any of his guests was unhappy, he would listen and say, "What do you want me to do? You tell me what you want me to do, and I will do it. If you don't want to pay the bill, that's fine. If you want to pay a part of the money, that is fine. Whatever you want." Oh boy! Wasn't that a disarming negotiation! He also did that with his good friends during the hard time. Particularly during the war—World War II—when people could not afford what they could before, he would write them letters saying, "If you cannot pay, that is fine. Just stay. If you have money later, you know what the bill is, and I trust you will pay it." (Interview with Doll, 2009)

The diplomacy that Doll observed as a schoolboy certainly played a role in his later leadership experiences at different levels. By offering flexibility to the other party while still honoring responsibility, Doll has been able to get things through in a patient yet persistent way, as earlier chapters have already demonstrated. Interestingly, even though he was not in the "in crowd" at school, he did not try to blend in; nor did he try to rebel. He retreated from an unwelcome place and created his own space rather than being trapped by or breaking from the system that did not work for him.

Now as an internationally renowned curriculum scholar, Doll believes that he has managed to be himself as a scholar without associating with any particular camp, and he does not approach such non-belonging in a negative way. He feels that his scholarship is not part of the mainstream, but neither is it part of the radical. He has worked out his own pathways that do not necessarily coincide with others'. Although he has been considered a postmodern curriculum scholar, he does not feel he belongs to the postmodern group. Most scholars using postmodern frameworks have come from political or humanity frameworks, but he quickly connected postmodernism with fractals and new sciences. The "odd" combination he put together was not received warmly but, instead, was often met with silence in academic circles.

Furthermore, as a joyful person by nature, his version of postmodernism is often in contrast with the pessimistic, language game of post-structuralism. We as students fondly called William Doll "an optimistic postmodernist." Triche goes further to say that Doll's postmodern spirit is also embedded in a search for the premodern "spiritual, sacred, and mysterious" (interview with Triche, 2012). Rejecting linear time, Doll's unique brand of postmodernism may very well be marked by such a blending of temporality. In this non-belonging space, Doll welcomes being "out of place." He is at peace with being different.

An Educator's Peace

> Control is present but it is not imposed, it emerges from the interactions present. (Doll, 1998b, p. 313)

Pedagogy in Flow

Doll discusses his first experiences of being a teacher in a private school:

> At school, a big boy of my size held a gun and a knife—I tried to let him give me the gun and knife. I had to take care of everything for students. Later when I was in the hotel business, I also had to deal with folks who were drunken. It is from such experiences that I have learned firmness and calmness in negotiation. (Interview with Doll, 2009)

The combination of firmness and calmness is the quality of peace in its double role of setting up the boundary against the action of violence and peacefully negotiating a productive space for involved participants to deal with conflicts. Firmness and calmness are the double gestures of nonviolent engagement (Slattery, Butigan, Pelicaric, & Preston-Pile, 2005; Wang, 2014). Such a combination is the foundation of Doll's recursive efforts to enact a pedagogy of peace in the classroom and in public, especially when turbulence and perturbation set in.

Doll's peace with difference is clearly demonstrated in the classroom. When there are contesting viewpoints among students, he patiently waits for all students to fully discuss the issue rather than jumping in to redirect the discussion in a more comfortable direction. He does not find disagreement an issue to be resolved but approaches it as generative for fresh ideas. Suspending his own judgment, he offers a space for students to fully explore. He explains his pedagogical creed:

> In a reflective relationship between teacher and student, the teacher does not ask the student to accept the teacher's authority; rather, the teacher asks the student to

suspend disbelief in that authority, to join with the teacher in inquiry, into that which the student is experiencing. (Doll, 1993, p. 160; italics in the original).

During the process of inquiry, he listens to students attentively and reflects their own ideas and thoughts back to them in a different way. This difference offers another lens through which to reexamine the issue at hand. There is a sense of flow in discussions among different viewpoints.

When I was a student, I was always amazed by his ability to draw the best potential from each student and navigate among students' different perspectives. His in-depth engagement with students is also coupled with his non-attachment to any specific outcome for students. I term such teaching "a pedagogy of nonviolence," teaching in which engagement is not instrumental but existential (Wang, 2014). Engagement without attachment means not pursuing a predetermined destination but working from what emerges from the process. Here flow, peace, and nonviolence come together through engaging difference.

Hendry offers another insightful reading of Doll's pedagogy regarding the flow of the class. She found his faith in the process fascinating, and at times, also perplexing: "Why are we teaching? What are students supposed to get? Would students need to understand something in a certain way?" (interview with Hendry, 2010). They were teaching questions for Hendry but not for Doll. Alternatively, he trusted that teaching is an emergent process and he has faith that students will come out of the process with their own brilliance:

> It is not about getting it [knowledge]. Wherever you go, that's where you are supposed to go and it's okay. And that is the beauty, and that's what I've learned from Bill: To let go. You cannot force ideas, and you cannot force thinking. . . . I never had the feeling that Bill had an agenda. There was nothing to get, and the more divergent that the class became, the more excited Bill got. There is no knowledge that is connected to that. . . . It is a real art that he has. It is an art to truly let go and just let a process emerge and take its own path, which means giving up control. (Interview with Hendry, 2010)

Letting go is a position that a teacher is usually afraid to take, but giving up control of a structured curriculum to let things emerge and letting the class follow its own pathway can be empowering for both teacher and students. Knowledge itself is not a major concern to Doll, but the process of understanding, knowing, and being in its flow is at the center of his pedagogy. This approach does not fit into the history of standards as a way of improving teacher education for the past century, a history that Roy A. Edelfelt and

James D. Raths (1999) briefly narrate, and the contemporary standardization. It is not that the standard is not important for Doll—he has worked hard to improve the rigor of teacher education programs—but he is skeptical about any predetermined standard imposed from outside.

Enacting pedagogy in flow, however, does require a sense of belief that something good will happen as the result of the process. As Pinar points out, if Doll's teenage years were marked by mobilization, that experience might have influenced his strong intellectual and pedagogical interest in dynamics, because a dynamic process characterized his finding his way in the world (interview with Pinar, 2012). Doll has a strong faith that students can emerge from their struggles stronger, just as he was able to do when he was young. Metaphorically, flow does not stop at bumps but becomes more dynamic in going over or around obstacles.

Although Doll does not lay out a rigid foundation of knowledge in his courses for students to grasp, the inquiry process that his pedagogy of peace enables does enrich and deepen students' understanding of curriculum, pedagogy, and education as the result of their own process of grappling with knowledge in a personal sense. Donna Trueit comments, "One of the joys of co-teaching with him is his classical background and real breadth and depth of knowledge. He is comfortable with answering all questions . . . and giving a good response to a student" (interview with Trueit, 2010). So Doll's own knowledge base provides rich soil to help students grow their own ideas in communal inquiry. Students understand that knowing cannot be separate from the complexity of personal growth in a social and cultural context.

Hendry also points out that in teaching, Doll is always in the moment. He continues the class conversation by sending students emails during the week to keep the thinking process going. The class conversation never ends. In a sense, "he never steps outside" of the process of teaching that continuously unfolds. It is not possible for him to reflect on his teaching from a distance, as he is always in the middle of it (interview with Hendry, 2010). Teaching cannot be an object to study in its usual sense, just as subjectivity cannot be an object to study, because he is always in the middle of it, undergoing, doing, and experiencing. Doll's pedagogy of peace offers a safe and engaging learning environment for students, not so much through his conscious efforts to make students feel safe, but because his teaching flows through the class and students are carried through it. As they undergo the process, they become more confident in their learning.

In this particular way, teaching and learning happen in a web-like manner rather than individually. If Doll is always in the moment of improvising his pedagogical response to the situation, then reflection is already embedded within the process rather than outside of it because there is no "afterward." The teacher's subjective positioning and repositioning happen instantly to incorporate the feedback quickly in the process. That is probably one reason that Doll seems to always be so pleased with teaching; for him, it is an improvised and spontaneous movement and performance. In this sense, Doll is united with teaching.

While peace education literature affirms the importance of inner peace for pursuing outer peace (Lin et al., 2008; Oxford, 2014; Quinn, 2014; Wang, 2014), Doll's pedagogical peace is based on acting on the world through the positioning of the self in the system. Being at one with teaching, the subjectivity that necessarily requires a certain distance from the situation is not a central issue for him, and his inner peace exists *in his interaction* with the world. According to Pinar, Doll has formulated theories about how people think, how they develop, and what a dynamic curriculum is. As a result, subjectivity is irrelevant for him. In addition, generationally speaking, "this 'I' stuff really broke out in the 1960s and 1970s in the U.S." (interview with Pinar, 2012), but not during the time when Doll was coming of age.

Those major thinkers such as Piaget and Bruner whom Doll draws on have systematic theories. Chaos and complexity theory is also a systematic approach; so is Bateson's ecological view. Certainly, Doll picks up what is most dynamic about each theory, rather than any rigid developmental approach. Tracing Doll's intellectual lineage, Trueit draws a "curvy line" between his intellectual interest and his schooling experiences. What is most interesting in the theories of Dewey, Piaget, Bruner, and Prigogine is "the idea of transformation and change and a different sense of order" (interview with Trueit, 2010). Those ideas have helped him to imagine a different schooling, one that is less constraining and more open to flexibility, creativity, and the humanness of students. However, a systematic viewpoint, albeit dynamic, makes it necessary to minimize the role of an individual component, including individual subjectivity.

Not paying close attention to the role of subjectivity is also related to Doll's ability not to take himself too seriously. As Wendy Kohli recalls, Doll is humble and does not have many ego issues (interview with Kohli, 2014) even though he is "a larger-than-life figure." Pinar also comments that Doll is unique in that he does not care much about his reputation or career promotion but has focused instead on his students (interview with Pinar, 2012).

Humility is an important element of Doll's sense of spirituality (Doll, 2012). When the defensive mechanism of the ego is not strong, he can see a bigger picture. In addition, his accepting the humorous unveiling of his privileges and his self-mocking comments become ice-breakers in the classroom. His pedagogy in flow is enabled by the lack of central authority in the classroom.

Doll's lack of concern with subjectivity has its limits, as we have seen in different chapters. Pinar points out: "It is part of his personality but also his thinking. This formulated quality comes from his system view. It has its price." He quickly adds, "Too much concern with subjectivity can reveal insensitiveness to others [as well]. . . . I would fiercely defend him" (interview with Pinar, 2012). Those very limitations Doll may have also contribute to making his teaching extraordinary. He also encourages his students to have a playful approach to the self and the world and not take ideas or themselves too seriously. Bridging separateness to achieve peaceful relationships within the self and with the other, Doll achieves an inclusive pedagogy, an emergent curriculum, and a generative education through a sense of creative flow.

Integrative Effects of Teaching at Peace

A pedagogy of peace has an integrative effect on students. Doll's own journey of making peace with trauma, loss, and guilt throughout his life, his patience with students who go through various difficulties, and his compassion for those who suffer make it possible for him to connect with students and bring them a certain sense of peace. Alcazar was able to share some painful experiences with him as they established a trusting relationship:

> I was able to talk with him about my own struggles in the Philippines including the tortures that I suffered from. In fact, it was actually the first time when I could talk to someone about those experiences. For a while, it was too painful to talk about it. (Interview with Alcazar, 2012)

Doll was appalled by what had happened and said that he would never understand what Alcazar went through. This acknowledgement of the other's difference without imposing a posture of empathy to erase the other's unique experiences was reassuring. For Mikhail Bakhtin, empathy as merging with the other's position or suffering runs the risk of imposing one's own ideas and emotions on another person (Morson & Emerson, 1990). Dialogic understanding requires a certain distance and allows the other to exist as the other.

Alcazar's sharing happened during a time when Doll was going through one of his own major losses, creating a sense of sharing each other's grief but without reducing it to the same. Without Doll's genuine interest in a student's life and his respect and compassion for another human being, Alcazar might not have told his stories. In being able to talk to others about those painful experiences, Alcazar built the Kristevian connections between the semiotic and the symbolic (Kristeva, 2002) to create meanings. It was a transformative experience for both Doll and Alcazar. While Doll was not able to share words with others about his early losses when they happened, he made bridges for a student to feel a moment of peace in articulating pain and to transform it.

In general, Doll's style of teaching is peaceful with its pace and flexibility, giving students room to move around. Triche thinks that Doll "is a quiet teacher" who can dissolve students' "noisiness." He explains:

> I think Bill's desire to bring his students into the conversation more completely is more than any other professor I have experienced. He taught me a lot about the role of students in curriculum development so that the professor is not the sole expert. In his ability to allow the class to become, as he said, to emerge through his lack of parental control, he never loses control either. . . . It is a teaching at peace. (Interview with Triche, 2012)

A teaching at peace is simultaneous with a teacher at peace who embraces an emergent sense of control (Doll, 1998b). As Triche observes, Doll does not lose control *per se*. What he practices is a new view of control that is "neither externally imposed nor internally developed" (Doll, 1998b, p. 296) but exercised in the dynamic interactions at multiple levels. It is control that is open to change. The combination of change and control, along with the mixture of stability and flexibility, makes his pedagogy of peace dynamic and complex.

Triche further discusses the influence of Doll's teaching on his own teaching and how he has made efforts to "become a quieter teacher," one who engages in conversations without offering immediate answers, even though "it is clearly not my personality and I had to struggle with it" (interview with Triche, 2012). Triche loves intellectual stimulation and voicing his ideas but he makes a conscious effort in the classroom to allow pregnant pauses, go around the point to provoke thinking, and give space for students to play with readings. Triche also allows students to discuss their daily frustrations with administration because these students need a place to talk about them in order to hear the calling of teaching as vocation (Huebner, 1999). Such an influence on Triche's teaching does not necessarily come from Doll's direct teaching but from their pedagogical relationships.

A student from Russia, Anastasia Chebakova, discussed with me a class on writing research and deepening inquiry, co-taught by Doll and Trueit at the University of Victoria. This class was designed to guide graduate students from different disciplines to engage in interdisciplinary conversations through reflecting on their writing, deepening their understanding of inquiry, and communicating across difference. Majoring in political science, Chebakova was pleasantly surprised to find that Doll and Trueit taught differently from political science professors and Russian professors.

What struck Chebakova at the beginning of the class was "a very welcoming, warm atmosphere in the class in which nobody is afraid to talk" (interview with Chebakova, 2010). She did not know how Doll and Trueit created such an effect, but she felt liberated from the constraints of traditional education. For somebody who came from a different country and a different discipline as she did, she felt that her way of thinking, feeling, being was well-respected by her professors and classmates. Everybody sharing ideas in the class created a spirit of working together in peace. Chebakova appreciated that both professors worked with her to engage, support, and advance her thinking; this kind of engagement was something that she had not found anywhere else. Although she did not use the language of peace, liberating students from the expert or authoritarian teaching mode, having faith in and providing support to students as they work through difficulty, and fostering democratic conversations and sharing in the classroom are all related to a pedagogy of peace.

The instructors' modeling in the format of team teaching also facilitated students' collaboration among themselves and making peace with tensions. Chebakova shared a story of working with her classmate from a different country and a different discipline. They had mutual understanding despite difference. Once their disagreement caused conflict that led to uncomfortable feelings. However, Chebakova realized that the difference also helped her to better formulate her ideas. They resumed working together and remained friends after the class ended. Doll's pedagogy of peace helps students open up conversations and engage in meaningful learning and also embodies a productive approach to tensions and disagreement to encourage students to engage peace as an active process of negotiating with differences.

The following teaching episode highlights Doll as a quiet teacher who allows the voices of students to come forward clearly.

(Donna Trueit was working with a student, Rasunah, on her paper and making various suggestions regarding her writing and format.)

Doll: May I comment that as papers come in, we will send papers to everybody. I think the concept of listening you have developed is a very important point in regard to the whole notion of pedagogy: how we listen, how well we listen. Often as teachers, we think we have all the knowledge that everybody would need, so we don't listen to others.

Rasunah: In listening cultures, listening goes from birth and on, and does not stop, so listening is in the culture.

Doll: Fascinating! We are not a listening culture in the mainstream ideology. We are a talking culture, talking heads, haha [*laughs softly*].

Rasunah: But dialogue is a listening culture. That is a wonderful thing about coming to this class.

Doll: Oh, Yeah!! . . . Of course we need to listen not only to humans, but to all of the sounds around us, and even to the silences.

Jung-Hoon: To me, as we listen to others and the world, we also need to listen to ourselves.

Doll: Yes! It is a nice point!

Rasunah: I can give an explanation of our Nation's point of view if you want. We have different kinds of listening. I will just give you an idea of what it includes. In the First Nation cultures, in my culture, we have many different ways of listening. One is *Bzindamowin*. It is a way of learning that comes from listening to the cultural stories. And there is *Aadizookaan*. It is the listening to the spirit of stories . . . .

Doll: Wow! [*Doll made sounds of "Oh" to show his awe throughout Rasunah's explanation.*]

Rasunah: The *Aadizookaan* has existed for a long time and has already been made available for social interactions, and in some cases the *Aadizookaan* can communicate directly with individuals through a dream path. In the dreams, spoken words are replaced by the implicit meanings of thought. That is already in the culture and we understand that.

Trueit: Oh!

Rasunah: Another one is *Kendaaswin*, traditional knowledge from one generation to the next generation and to the next. Then *Manidoo-Waaniwin*, revealed knowledge, something that is revealed to you. And *Manidoo-minjimendamowin*, which means spirit memory, so you have a memory of your ancestors' experiences, which you can access, or you can say your blood memory. Those are just some examples of different types of listening.

[*The class members were talking all at the same time, excited by all these notions.*]

Patricia: Is there any publication on spiritual listening?

Rasunah: Yes, there is. It is already published. It is on the website.

Jung-Hoon: Can you send us the link to the website?

Rasunah: Sure. (Video transcription, April 13, 2013)

Here the capacity for listening became both the content and means of discussion. A particular student's contributions to understanding what listening means enriched everybody's understanding as they listened to one another peacefully—there was no competition but receptiveness and eagerness to learn from different modes of listening. The indigenous ways of listening and the necessity for listening to both the self and the other, both the human and the ecological, and both sound and silence were central topics of the discussion, all of which are essential elements for co-creating a pedagogy of peace. A class community-in-the-making supports such a pedagogy as everybody participates in a peaceful culture that nurtures intellectual and spiritual explorations. "The interplay between the individual and the communal" develops "a network of connected and interconnected thoughts" (Doll, 2008/2012, p. 30) in which thinking at peace is intertwined with a passion for new knowledge. Doll showed his enthusiasm for the notion of a listening culture in various verbal and nonverbal ways. His pedagogy of passion and pedagogy of peace come together—peace keeps passion balanced and passion keeps peace alive.

I have not found much literature in teacher education regarding a pedagogy of peace, but it is important today, when human conflicts and ecological disasters that threaten humanity and the cosmos are on the rise. When a teacher does not embody peace, peace cannot be taught to students. The marginalization of peace and nonviolence in the content, method, and purpose of teacher education, especially in today's climate of accountability and standardization, must be disrupted by teacher educators. Doll's pedagogy of peace enacts a proactive and creative stance. As Sandra Wasko-Flood (2010) argues, "Creativity and peace are intimately related: the student who discovers his or her creativity is more likely to find peace" (p. 146). Peace, creativity, and passion come together in the fluctuation of life, teaching, and learning to co-create a better world.

In today's American education, the win-or-lose mentality is the mainstream with the triumph of a testing culture, a mentality that causes the malnutrition of students physically, intellectually, emotionally, and spiritually. The win-or-lose mentality is absent in Doll's teaching; instead, the professor's and students' co-creative passions are directed to sharing ideas, experiences, and insights in a communal inquiry. Enacting emergent control, forming peaceful relationships with the self and the world, accepting the positive role of tensions, and creating collaborative teaching partnerships all contribute to Doll's pedagogy of peace. In this dynamic peace, everyone's participation becomes crucial, to which I turn in the next chapter.

NOTE

1. William Doll suffered a huge loss when he was divorced from Mary Aswell Doll—at the time they had been married for more than twenty years. This book's focus is on his pedagogy, so I do not discuss such a personal issue. Although it was traumatic when it happened, over an extended period of time William Doll has learned valuable lessons about relationships.

· 7 ·

PEDAGOGY OF PARTICIPATION

> The "essence" of our being, if I may borrow the metaphysician's concept, is to be dialogical—to have interaction with others in a community. Goals, plans, purposes, procedures, judgments, evaluations all come from this sense of community. (Doll, 1990/2012, p. 149)

One landmark of William Doll's pedagogy is his ability to build a community. Trueit told me this story about her first experience of his teaching a summer class at the University of Victoria:

> I was taking a different class that semester, and Bill's class was just before my class. When I waited outside of the classroom for my own class, I could feel the difference that his class brought to students. In the first few days, students came out one by one, then in the next couple of days they would come out talking to each other animatedly in twos and threes, and by the end of the course, we had to knock on the door to get them out of the room. They would come out *en masse* and would be talking to each other loudly going down the staircase: they had coalesced into this group, and it was no longer twos or threes but the whole group walking with Bill down the stairway, talking. It is pretty exciting for students of curriculum when you have to drag the class out of the classroom as they were so engaged with the class. (Interview with Trueit, 2010)

Doll is adept at building a class community in which the relationships are not only between the teacher and students but also among students. It is not so much about using any particular techniques, but about his capacity to connect with others and help others to connect. Everybody participates in his classes. His pedagogy of participation has not been limited to the classroom but extended to local, national, and international communities. What underlies this philosophy of participation is a strong sense of responsibility—political, ethical, and ultimately spiritual responsibility.

Philosophy of Participation in Teaching and Leadership

> Development is not continual and gradual; it is punctuated with plateaus, spurts, and bifurcation points. It also requires the cooperative efforts of a community (class). Development occurs more through cooperation than competition; although each has its place and role. (Doll, 1986/2012, p. 142)

Doll, engaging in what Triche calls a "philosophy of participation" that encourages "everybody to be in the circle" (interview with Triche, 2012), has an amazing ability to bring everybody to the conversation and make multilayered and recursive connections between and among participants and ideas. In *Democracy and Education*, Dewey (1916/1997) points out, democracy is "more than a form of government"; it "is primarily a mode of associated living, of conjoint communicated experience" (p. 87). Doll's pedagogical orientation is participatory democracy. Whether through seminar discussions, small group activities, or individual conversations, he interacts with every student and encourages their participation.

After many years of activist work, Alcazar returned to the university without knowing where he would fit. However, Doll made him "feel welcome, at home" when he first walked into the classroom: "He arranged the seats like a circle. That, for me, was reassuring already. Because I was really rebelling against the traditional seating arrangement in which the teacher is in the location of power in front of the class" (interview with Alcazar, 2012). In Doll's classes, Alcazar was introduced to Michel Serres's book *Troubadour of Knowledge* (1991/1997), which discusses the notion of the instructed third, or the third excluded. The notion of going beyond dichotomy and finding alternatives spoke to Alcazar's activism. So the classroom setting and the readings, as well as having lunch together as a class, appealed to Alcazar, who built meaningful bridges between his social and political activities and intellectual discussions in the classroom.

Doll sees the responsibility of a teacher educator as:

> Helping students develop a sense of meaningful participation in society in which they live. By meaningful participation, I mean not only [for students] to be active but to understand the roots of their society well so that they don't get caught up in the immediate issues but rather see that issue in historical and multi-focused perspectives. (Interview with Doll, 2010)

Both the intellectual history of ideas and multiple perspectives of the same issue are important for Doll's postmodern pedagogy and curriculum approach. He has in-depth knowledge about classical thought in Western philosophy, and in his scholarship he often traces the historical evolution of a certain concept or a certain idea in order to reveal alternatives.

Doll (2002c/2012, 2002d/2012) discusses the historical legacy of Peter Ramus and the Ramist map in the sixteenth century and its impact on the contemporary methods courses in teacher education. Making connections between the Ramus map, Calvinism, industrialization, the scientific management movement, and the Tyler rationale in curriculum studies, Doll presents a broad picture of how mainstream education as a field has been shaped. Adopting a historical perspective, Doll believes, is essential for enlightening contemporary education and finding alternative pathways because history reveals the conditions and contexts for making one particular pathway possible. Then the question becomes, How does one go about recreating different conditions to seek alternative pathways? Students in teacher education as participants in a democratic society must be historically informed.

In choosing texts for students to read, Doll often blends classical with contemporary works. As I discuss in Chapter 4, in a mathematics discipline course he regularly taught at LSU, he chose *Flatland*, an intriguing novel, published in 1884. Who would imagine that such a novel would be included in a mathematics education class? However, this book served the purpose of developing students' meaningful participation in society because their mathematical understanding is combined with social and cultural critique. Not constrained by but going beyond their own thought processes, students can release mathematical and social imaginings to delve more deeply than ever before.

Multiple perspectives are always an important part of Doll's pedagogy. Without this openness to different perspectives, it would not be possible for every student to feel free to participate. He organizes the class in such a way that each student has an opportunity to make connections with their own teaching or life experiences, which in turn opens up multilayered conversations. Rarely have Doll's classes been dominated by one or two students, as often happens in other classes. Somehow the class dynamics dissolve the emergence of one center and students respect one another's space for participation. While Doll seldom uses the format of students taking turns to speak, he is mindful of those who are not part of class discussions and improvises ways of inviting quieter students to speak up.

> When I was a doctoral student participating in Dr. Doll's class, I made my best effort to participate when I could. It was difficult in the beginning years, since I was not familiar with the context of intellectual ideas and I did not share the same historical and cultural framework with my professor and classmates. Everybody else seemed to know what was being discussed and gave their input while sometimes I felt like I was in the middle of nowhere, and I was not sure how to insert my thoughts into the flow of conversation. In my own cultural context, listening is a good way of learning if one listens well and takes things in. I listened to the conversation a great deal. There was one time when a guest speaker came in to discuss the influence of the Puritan heritage on American education. I did not have much knowledge about Puritanism so I simply listened to the presentation and discussion.
>
> Approaching the end of the class, Dr. Doll asked: "Could those who have not spoken yet speak?" I was one of two students who had not spoken yet. Being called upon in this way drew out my intellectual aggressiveness, I guess, so I responded by saying that the legacy of the Puritan heritage around the world was American superiority because Puritans were the ones chosen by God. Well, that was not the expected response. Although not always pleasant, Dr. Doll's expectation of everybody participating in the class has helped me, over time, to develop intellectual and social skills to communicate and exchange ideas with others, a basic skill to survive in American academia. In my own teaching after I became a teacher educator, I have also made sure that all students have chances to participate in class.

As Emily Robertson (2008) argues,

> [T]he knowledge, skills, and virtues required to participate in democratic practices of engagement set goals for the education of democratic citizens, as well as supplying goals for the education of the teachers who will help develop these capacities in each new generation. (pp. 29–30)

Such democratic practices of engagement and participation—embraced by students—mark Doll's classes. While Robertson focuses on social and civic engagement, pre-service and in-service teachers also enjoy intellectual engagement in Doll's teaching. They practice and foster the willingness to listen to others' points of view and "change [their] views in the give and take of conversation" (Robertson, p. 29). They engage in discussions of controversial issues and conduct critical communal inquiry on shared topics of interest. Robertson terms such capacities the virtues of deliberation, and clearly Doll's students deliberatively converse and discuss educational issues. But they do more than that if deliberation is defined as a process of "rational persuasion" (p. 40) because they bring stories, emotions, and spirituality into the classroom and their participation is not merely rational but a whole-being experience.

Unlike some professors who choose to remain somewhat distant, Doll always participates in students' small group activities in the classroom. As a teaching assistant in an elementary school curriculum class he taught at LSU, I witnessed how he organized small groups. At the beginning of that class, Doll divided students into small groups and these groups stayed together for the whole semester to engage in all kinds of activities. While the formation of the groups was random, the group members developed close relationships by the end of the class. For Doll, building a community that encourages democratic participation requires sustained engagement, so he created conditions for temporal sustainability in small group activities, in which students had more opportunities to share their ideas.

When it was small group discussion time during class, Doll stayed with one group for five to ten minutes, listening to the group conversation very closely, asking questions when needed, and giving advice if requested. Then he moved to another small group and picked up things in the middle of the conversation to make his comments. I was amazed by his ability to move among groups so easily and capture important points while giving his input. He visited all small groups before the end of the class. With increased participation and the professor's individualized and contextualized guidance in small group discussions, each student became more engaged with their learning.

Doll also makes sure to have individual interactions with students, sometimes setting up an appointment to meet each student during the semester, sometimes including individual consultation time within the class, and always inviting every student to participate in gatherings. Anastasia Chebakova appreciated the individual help with her thinking and writing, and she was amazed by Doll's and Trueit's ability to engage her thinking, understand what she was going through, and advance her ideas. At one point, she tried to start a project and felt frustrated; she then went to Doll for advice, and within five minutes of their conversation the problem she was dealing with was resolved (interview with Chebakova, 2010).

International Teaching

Doll taught at the University of Victoria in Canada during the summer for a number of years. When I was his graduate assistant, I helped him send international priority or express mail at his own expense to his Canadian students before a class began. I remember my skepticism, and I asked him, "Do these students actually read before the class begins?" He said, "They may or may not, but often they do. It is an opportunity offered to them." He often sent out these materials with his love notes, just as he often did for American students. Generously walking the extra mile, he must have inspired more students to read and think before the class started. It is an invitation, and while many students respond, Doll does not expect everyone to do so. He creates conditions for more students to participate, but he is not attached to any particular outcome.

As I discuss in Chapter 5, Doll did not change his open discussion format in his international teaching in China. He did not change his philosophy of participation in teaching, but he did improvise strategies to make meaningful connections with Chinese students. For example, when Doll and a group of professors and graduate students were on an academic tour of different Chinese universities in 2007, they came up with an arrangement that encouraged Chinese students to participate in conversations. In addition to translators, there were four American graduate students, so Doll asked them to sit apart from one another in the middle of the Chinese audience that included many graduate students. American professors and translators were on the front stage.

Because the audience was a large one, even if American professors did not want to reinforce a hierarchical pedagogical system, their invitation to Chinese students to participate was not well-received because of the constraint caused by differences in traditions and languages. Even though the Chinese students did not want to talk to American professors directly either in English or in Chinese, they did turn to American students to talk to them in English, albeit halting English. Then the American students told the American professors what the Chinese students said, and the professors came back with more explanations and ideas, which were translated into Chinese. The direct communication between American students and Chinese students without the mediation of linguistic translation added a rich layer of intercultural conversation that would not have been possible in a linear teacher–student relationship.

In a broad sense, translation is about connecting across differences in a two-way process, in which all participants undergo a certain level of change after their encounter (Wang & Hoyt, 2007). Jacques Derrida (1987/1991) points out both the necessity and impossibility of translation and understands it as "involving the same risk and chance as the poem" (p. 276). Translation is a creative act, living with the tension between the self and the other poetically. Ted Aoki (2005) plays with English, Japanese, and Chinese languages to demonstrate how the translation of language and culture generates new pedagogical meanings. In this sense, international teaching is a form of translation, with or without language differences.

Despite the educational tradition in China, conditions can be created to encourage more participation from students, as young people of any culture seek opportunities to become themselves despite political and cultural constraints. The dissipative control (Doll, 1998b, 2012) that was enacted through American students at multiple loci of conversations freed Chinese students from a formal relationship with the teacher. Intercultural and linguistic translation happened on multiple dimensions.

Recent Chinese curriculum reform has also aimed at opening up more possibilities for the younger generation to participate in more meaningful ways and become creators rather than followers (Chen, 2014; Pinar, 2014). Good translation can contribute to such a process. Part of the appeal of Doll's pedagogy to the Chinese audience is that his participatory teaching orientation matched the need for educational and cultural transformation in China at this particular time. He and his colleagues used that particular format in different universities and found it effective in various settings to engage in conversations, multilayered, bilingual, and intercultural conversations.

Team Teaching

Team teaching is part of Doll's pedagogy of participation. Since he first collaborated with a colleague in Denver to run the Great Books program, he has proactively searched for co-teaching opportunities and informal collaboration. At SUNY Oswego, he worked with two other professors who were also teaching a social foundations class. They did not necessarily use the same books in their classes, but they met every two weeks to talk about what they were doing. While informal, their conversations influenced their teaching in their own classrooms. At LSU, Doll was comfortable with any other professor stepping into his class, and some colleagues did stop by to participate in class conversations.

Fred Korthagen, John Loughran, and Tom Russell (2006) point out that an important principle of teacher education is "an emphasis on those learning to teach [to work] closely with their peers" (p. 1032). Teaching can be an isolating profession if the teacher does not actively go beyond the walls of the classroom. Team teaching leads to improvements in practice as teachers share with and learn from each other (Loughran, 2006). Creating such a learning community for students in teacher education, however, also requires collaborative teaching among teacher educators so that students can experience the dynamics of team teaching as learners in order to gain insights into how to teach in collaborative ways.

I have discussed Doll's team teaching in earlier chapters, but here I would like to highlight the transformative power of team teaching in a pedagogy of participation. For Doll, participation leads to the transformation of both participants and the community through interactions. This approach is influenced by the Deweyan notion of reconstructing experience, Piaget's notion of interaction, Whitehead's process-oriented viewpoint, and self-organization theory (Doll, 1993, 2012). In team teaching, the instructors' respect for each other, their willingness to de-center authority roles in the classroom, and their ability to negotiate and play with difference demonstrate for students what collaborative participation and mutual learning can be.

The course on pragmatism and feminism co-taught by Doll and Hendry is a good example of mutual learning. As the class started, Doll knew little about feminism and Hendry was skeptical about pragmatism, as seen in her critique of the universal nature of "experience." Doll's critique of feminism was that one singular lens is often biased, whereas Hendry's critique of pragmatism was that it is too rational and less embodied. Specifically, she thought Dewey did not pay enough attention to the influence that Jane Addams had on him. By the end of the class, Doll had become more aware of women's role in pragmatism and education; in his own words, he began to "see things much more from a woman's point of view. I am aware how often I go to males naturally on this or that and I have to keep that in check. I just overlooked it" (interview with Doll, 2010). Hendry became more appreciative of Dewey's school of thought, especially his notion of action. However, she still thought that Doll did not see Jane Addams as a philosopher, only as a social philosopher.

Both Doll and Hendry got to know each other's teaching styles. Doll recalled her teaching with appreciation:

> Petra, in her teaching, always deferred to everybody in the room, not just me. She does not speak first. She really prefers to be the last speaker. And so in our seminar she always let me speak first. And then I started and then we had a conversation that went around. At some point, she would come in, articulate her own view, and related it to the other views that were circulating. She would ask us to think about it this way, and think about it that way, and she just raised the whole conversation to another level. (Interview with Doll, 2010)

In Doll's description, we can see that Hendry's teaching reflected a feminist pedagogy, emphasizing that everybody in the class has a voice and the feminist teacher seeks to create the conditions for all students—particularly those who are marginalized and do not have many opportunities to voice their own perspectives—to contribute to the learning process that is both individual and communal (hooks, 2000; Miller, 2005). As a post-structural feminist, Hendry challenges students to imagine gendered identity as fluid and complex but she is firmly grounded in creating a class atmosphere for promoting gender equity and social justice (Hendry, 2012). In her team teaching with Doll, she was not afraid to interrupt when she perceived the class dynamics as centering a comfortable male voice (see Teaching Scenario 5.1).

Teaching Scenario 5.1

William Doll and Petra Munro Hendry are co-teaching a class on pragmatism and feminism. Doll has talked for quite a while in the class.

Hendry: Bill, what about the other related notion we discussed earlier?
Doll: Oh, a good question to think about!
Hendry: How about we let students respond first?
(Students are encouraged to steer the conversation and some began to talk.)

Hendry also learned from Doll's teaching:

> I learned a lot from him by just watching him and how he teaches and he thinks very uniquely, very complexly. He looks at things very spatially with the math perspective and I'm much more the historian, so we complemented each other.... He does not see the world in categories [such as gender and race].... For him all things are related, and they are inter-relational and I think that partly comes from his math background. (Interview with Hendry, 2010)

Students were able to "see two different points of view that can work together to have a sort of healthy tension that becomes generative" (interview with

Doll, 2010). Everybody participated in such conversations, and everybody learned something from the process.

Team teaching with both Hendry and Trueit had helped Doll become more mindful of women's role in society, culture, philosophy, and spirituality. Both Hendry and Trueit were affirmative presences in team teaching and gently challenged his way of thinking to make him realize that the relational, even though not categorized, needs to be concretized in specific interactions that sometimes have a particular lens of gender, class, or race. Both Hendry and Trueit commented on his change over time, pointing out that he learned to let go of his control and became a better learner over the years.

Alcazar also noticed that Doll's language had become inclusive over time. In one of his recent classes at UBC, Doll commented, "As we think about curriculum, it has been in a mechanical sense. 'Curriculum' is a noun, is designed in a linear manner, *male-oriented* [approach]" (video transcription, March 9, 2013; italics added). Explicitly using the gendered language in his teaching, he has shifted somewhat from his previous positioning. The ongoing process of interacting with colleagues and students who are different from him and team teaching with women teacher educators who respond with their own sense of pedagogical authority have contributed to his transformation. In other words, Doll's philosophy of interactive, transformative, and democratic participation that leads to transformation has worked magic for his own change as well.

Leadership Experiences

Doll has assumed leadership roles on many occasions and he has encouraged participation from all parties whenever he could. When Doll was the headmaster at Valley School in the early 1960s, his participatory orientation invited all teachers to play important roles in the school. As Mary Aswell Doll recalls, "The atmosphere at his school [was] warm and happy: he had excellent rapport with his faculty, a group we partied with frequently" (personal communication, December 30, 2015). He encouraged both students' and parents' participation in various ways:

> I did not have a library room but I used the corridors of the building as the library. We just put books on the shelf. As students were passing from one room to another room, they would pass the books. They could bring books home. We really trusted that they would bring them back, and the book loss was not very great. If these books ended up in somebody's own library, that was fine. These were good books for kids.

And then I wanted to bring art in. We had an art gallery of practicing artists and the art products were for sale, located in and among the books, so the books would not go all the way up to the ceiling and there were paintings here and there. When parents came in, they saw real, live art, done by local practicing artists. And they saw all these wonderful books. (Interview with Doll, 2009)

Inviting students to read books by creating a school library in the corridors of the building, Doll was encouraging students to participate in conversing with books; losing a book or two along the way was unimportant. Not only did Doll invite local artists to place their work in the school, he also put on demonstrations of children's play with Cuisenaire Rods for their parents. He asked parents and children to sit down and work together with the rods, and the first-grade children showed their parents how to play in different ways. Many parents were intrigued and these parents' sense of connection with school was strengthened though participating in their children's education.

However, such engagement did not guarantee agreement, and encouraging parental participation did not mean yielding to the pressure of parents who wanted to impose their agenda on the school. Before he became the headmaster, Doll was the Director of Mathematics at the Graland Country Day School in Denver. He listened attentively to what students were saying, picked up on the conversations, and asked a few questions here and there. He also talked to parents about what he was doing. Most parents were supportive but a few held a different view and did not like his orientation toward free-thinking. He tried to deal with issues as gently as he could, making sure that the foundations of subjects were taught well, but he did not let those parents dictate how he could teach mathematics. Doll's notion of participation is to create the Deweyan democratic way of living in school without accepting closed-mindedness.

As the headmaster at Valley School, Doll also encouraged teachers to work together and found time for them to engage in collaborative teaching activities. In today's school environment, time is still an issue for collaborative work but even as the school leader, Doll participated in daily educational activities:

I would read to children at lunch, and I read them fairy tales. I wanted them to understand their culture and various other cultures, and I read them in a dramatic manner and of course that kept them quiet, so there were fewer food fights. They also got introduced to good literature at their age. (Interview with Doll, 2009)

Reading to schoolchildren on a daily basis at lunch, Doll enacted his leadership through directly participating in teaching students and inspiring their interest in reading and culture. He also ventured outside of the school to proactively connect with a nearby teacher education program:

> I went to Goucher College, a nearby college with a teacher education program, and asked them to place their student teachers with us, but I had one requirement, and that was that student teachers assigned to my school had to stay with us for a full year. The college had difficulty with this requirement, as their program was not set up in this way, but they did so for us, and they did so basically with those who were going to graduate school. (Interview with Doll, 2009)

Professional development schools had not yet started to appear on the scene, but Doll's vision was that establishing collaborative relationships between the teacher education program and the school was beneficial for both. Doll made sure that students who participated in internships at his school would stay for a full year so that engaged relationships, in which members worked for shared interests and worked out differences, could be established. At Valley School he made extensive connections with students, parents, teachers, and the local community and a college to create a learning community for his school. Collaborative participation has been the hallmark of Doll's leadership experiences. His pedagogical visions for teacher education were nurtured by all these experiences. Individualistic ways of learning, either at school or in teacher education, have never been his preference, and he has always searched for ways to encourage students' participatory and collaborative learning.

Creating a Community of Learners

> In this post-modern act we lose the certainty modernism posited but open ourselves to a sense of community it never had—a dialogic community (Doll, 1990/2012, p. 144)

As mentioned in Chapter 1, Friday Friends lunch gatherings became a tradition while Doll was teaching at LSU, and many LSU graduates remember them fondly. He taught doctoral seminars for a number of years on Friday morning. After the class, he invited all class members to have lunch together. Sometimes a couple of colleagues joined in the lunch—David Kirshner was one who frequently did so—and conversations were engaging—intellectually, socially, and personally—and full of laughter. The topics of conversations in class often got carried over to lunch. Organizing such a gathering, however, had been part of Doll's life long before he came to LSU.

When he was the headmaster at Valley School, he was also enrolled in graduate courses and hosted monthly seminars in his rented room, inviting professors, graduate students, and teachers to come over and discuss educational issues. Professors and students also presented papers that they were currently working on.

> As a graduate student I had the opportunity to come in contact—in a wonderfully conversational and friendly manner—with a side of behaviorism (Skinnerian) I had previously shunned. While I did not adopt this view, too rigid and manipulative for me, I did begin to acquire the art of listening to and negotiating with those of another persuasion. This skill has served me well for over 40 years, as both a teacher and administrator. (Doll, 2009/2012, p. 16)

Being able to engage others who hold different perspectives is a craft that he honed, and patience with others is a quality that he developed, even though Doll sees himself as "an organizer by nature" (p. 17). This "art of listening to and negotiating with those of another persuasion" is important for creating a community, "a community without consensus" as Janet Miller (2010) envisions. When a community is marked by sameness, it is no longer dynamic; for Doll, keeping knowledge, relationship, and community alive and open to difference is crucial.

At SUNY Oswego, Doll, along with his colleagues, started a Piaget reading group that gathered regularly in his house to discuss Piaget's writings. At least six teachers participated in those discussions. During that time, Doll formed his own perspective on Piaget's major work. At the University of Redlands in California, he gathered a group of scholars from different places, including faculty at his university, colleagues at Claremont Process Theology Center, and professors at California State University at Fulton. At the time, Doll was reading Whitehead and Prigogine, and his postmodern framework came much from his interactions with David Griffin, as well as the work of Daniel Bell and of Charles Jenks.

The groups that Doll organized over the years were usually interdisciplinary, and people from diverse disciplines and backgrounds came together to explore issues. Creating a participatory culture in which people share ideas and food characterizes Doll's pedagogy of participation in a broad sense, not only as a teacher and teacher educator but also as an intellectual and a citizen in a democratic society.

Friday Friends also appeared in another form at the Bergamo Conference on Curriculum Theory and Classroom Practice[1] after Doll joined the group and attended its annual conference in the 1980s. Doll did not feel that he was established as a scholar compared with his peers in the group, at least not until he published a major book in 1993. He did not know the post-structural jargon and was not involved in critical theory. Feeling like an outsider at the Bergamo conference, however, did not prevent him from setting up an alternative framework for intellectual conversations. On Saturday morning, he asked for two sessions' time so that different perspectives on a similar topic could play off one another.

What he had not seen at Bergamo was a group of people sitting down to explore issues, so he asked to have a special session with a longer time frame in which people did not deliver papers but instead put out ideas for conversation. Dealing with his own sense of being an outsider at an established conference, he used the strategy of "retreat and create," similar to the one he had adopted when he was a teenager to open up spaces for intellectual freedom and exploration, and in so doing, he invited others to participate and seek alternative pathways. As Mary Aswell Doll points out, "Bill was always interested in finding solutions through various means of thinking, of discovering more than one path to an answer" (personal communication, December 30, 2015). The space for playing with ideas was expanded by those Saturday morning sessions.

Doll did something similar in founding, along with others, the Special Interest Group (SIG) on Chaos and Complexity Theory at the American Educational Research Association (AERA). Although the AERA conference has grown huge and impersonal, he ran the SIG in a manner that defied the typical AERA style. The leaders changed the presentation structure to open up conversations rather than asking presenters to rush off to a twelve-minute presentation. After the SIG business meeting, everybody was invited to go out to have dinner together. Just as team teaching leads to transformation of both teachers and students, participation in the academic culture also gives birth to new meanings. Doll has always been keen to create a more inclusive space for participatory conversations beyond institutional constraints. William Pinar comments that Doll has always been creative in dealing with the institutional aspect in order to open up intellectual space (interview with Pinar, 2012).

Dedicating one of his books to Doll, Pinar (2006) offers a testimonial to the impact of Doll's friendship and colleagueship:

Bill has always demonstrated a profound concern for me as a person, a keen and enduring interest in my career, and an unquestioned commitment to our common cause; a sophisticated field of curriculum studies through which the practice of education might become more complex, more nuanced, more progressive. That commitment has been a reflection of Bill's own pedagogical practice, a practice not confined to the classroom, but extended in highly individuated ways to those of us he has so generously befriended. (p. xv)

Doll's genuine interest in others' work and life has connected him with many who may agree or disagree with his perspectives in his pedagogy of participation both in and out of the classroom. In 2001, in honor of Doll's contribution to the field of curriculum studies locally, nationally, and internationally, Pinar hosted a conference, In Praise of the Post-Modern, at LSU to celebrate William Doll's seventieth birthday. Colleagues and former students from all over the country attended the conference and gave presentations on Doll's curriculum theory and its impact on their own work. The event itself demonstrated the ripple effect of Doll's pedagogy of participation.

Creating a community of scholars is Doll's biggest source of pride at LSU. In a graduate program, students usually take a series of courses, but don't "necessarily establish a relationship among those courses" (interview with Doll, 2010). Because many students also work full-time, it is difficult to share a sense of community. The Curriculum Theory Project at LSU, created by Doll and Pinar, became a site for professors and students to interact with one another and form "a community of scholars" (interview with Doll, 2010) that was inviting, warm, and inspiring. National and international speakers were invited to give formal and informal presentations to both faculty and students multiple times every year and parties were hosted at the end of the semester. These events served as both intellectual and social occasions in which ideas and relationships were enjoyed and shared in a playful spirit.

Because of these efforts, students had a group of faculty to go to regularly for intellectual conversations and a group of classmates to go to for discussions. As explained in Chapter 5, Curriculum Camp, a graduate student conference, was also established at LSU to mentor students. In addition, LSU hosted national and international conferences in 1999 and 2000, which initiated the internationalization of curriculum studies movement in the field (Pinar, 2003; Trueit, Doll, Wang, & Pinar, 2000). Being at the center of all these exciting intellectual events, graduate students became inspired and eager to participate.

Doll has taken pride in both the Holmes elementary program and the Curriculum Theory Project:

> These are things I had wanted to do all the way back to my first teaching days in Oswego. And that was in the early 1970s and they came to fruition in the early 1990s, twenty years later, and I was just overjoyed at being able to have played a part in this sort of development. (Interview with Doll, 2010)

He played a leadership role in bringing people together in his generosity to others (including financially supporting graduate students), his genuine interest and pride in students' growth, and his passion for intellectual ideas.

Doll collaborated well with Pinar, who had a "sense of gathering and giving people flexibility to be innovative" and had "many imaginative and wonderful ideas" (interview with Doll, 2010). Doll picked up some of Pinar's ideas and turned them into reality. As discussed, such efforts necessarily involve playing with the system for implementing good ideas. With the skills he learned in his youth and his extensive experience in community building, Doll skillfully negotiated an inclusive space for persons from diverse backgrounds to nurture a "mutually flourishing community" (Hershock, 2012) at LSU and beyond. David Kirshner (in this volume) also comments on such a community from a colleague's viewpoint.

Politics, Ethics, and Spirituality: Participation and Responsibility

> Education in the sense I am considering is education which focuses on *our being*—on our engagement with life as this is manifest in humanity, the world, the universe, and the cosmos. Such an education does struggle with the spiritual, and is infused with the spiritual at the same time it infuses the spiritual with us. This is an education which questions the being of all we hold sacred while at the same time manifests a faith that such questioning will lead us to the sacredness of being. (Doll, 2002b/2012, p. 42)

Doll has not discussed education and teaching in any categorical way and he has not been much concerned with the impact of race, gender, class, sexuality, or any other identity on education. However, he has been influenced by broader political contexts and has held leadership roles at various levels of institutional change. He has performed a balancing act according to what the situation called for but has not affiliated with any identity politics.

Both William Doll and Mary Aswell Doll took an active role and participated in the social changes in the 1960s:

> Mary and I went to a small African American community and registered African Americans for voting for the first time. Whites were furious about us. Blacks were just so amazed that they invited us to their special occasions involving their church. . . . [And] we did then begin to march. Those marches in Washington happened periodically. Busloads would come from all over the country, and in Baltimore, so close to Washington, loads of people came there in order to march in Washington to protest the war. Mary and I joined in that . . . .
>
> It was under Steve Mann's guidance—among a whole bunch of forces—that I began to think about my role as a citizen of this country and my sense of social responsibility. . . . Social responsibility does not mean merely following those in power but it does mean analyzing and thinking, and that sense of course has influenced all of my teaching. (Interview with Doll, 2009)

Participating in these activities has changed Doll's life and teaching, as I discuss in Chapter 2. I think Doll's political sense is part of his theory and practice of democratic participation that leads to transformation and change in the long run. However, Doll has a hard time situating his work in terms of politics, especially when politics is tied up with partisan politics, or identity politics, or "status politics" (Kliebard, 2004). He has remained an advocate of a multidimensional approach, of looking at things from a number of perspectives, because "I am very fearful of one right way" (interview with Doll, 2010). A skewed singular lens for looking at the situation has never found its way into Doll's thinking. But he does have a strong sense of the political, although that sense comes from working things out through conversation and participation rather than looking at the situation through the lens of identity.

As Pinar points out, Doll has a systematic view of education and teaching, influenced by thinkers such as Piaget and Bruner, who have a formulated theory of learning and growth. In such a systematic—a self-organizing system—approach, the notion of identity and subjectivity does not play an important role. Pinar says that he would "fiercely defend" Doll's intellectual brilliance and contribution to the field, even though Doll was not explicitly concerned with the issues of race or gender. Such concerns have become prevalent in education during the past several decades, but not without their own limitations (interview with Pinar, 2012).

Alcazar acknowledges that Doll's experiences were "in comfortable settings and situations," but adds that there were multiple forms of oppression, including oppression of ideology. Doll rejects all metanarratives no matter where they come from, so Alcazar considers Doll's approach as critiquing "the ideological part of [oppression], that mind-set that we are better than you, or anyone else. In that sense, he is very much in my field of social justice work" (interview with Alcazar, 2012). Alcazar also recalled that when Katrina hit New Orleans, Doll was one of the first people who communicated with him and one of the first to make a significant donation to Alcazar's project to collect wheelchairs so that the elderly and disabled could leave the city.

Doll has been influenced by Jacques Derrida's sense of respecting the otherness of the other as well as Richard Rorty's (1989) notion of liberal irony in keeping the conversation going. Doll's sense of the political is to work with others to make a better world, and "making a better world is to negotiate with others in a community" (interview with Doll, 2010). He has worked at the department, college, and university levels, participated in community service throughout his life, served on a school board with success, and worked with people not only from the local community but also around the world—for example, Canada, China, Finland, Japan, and Russia. He believes that every citizen has a responsibility to participate in community building.

His sense of participation is not following the rules or regulations, but being critically engaged in contributing to making a better world. His refusal to submit a "regular" syllabus is one example of his resistance to following the "rules." His leadership in changing the whole university's general education curriculum structure is a bigger example of negotiation with the system by not allowing the rigid order of a system to take over his initiative. He has also learned through experience that "when you are going to step out of line and do something, don't expect any support" (interview with Doll, 2009). When he, as the department chair, challenged the university's authority on behalf of a colleague, all other colleagues, out of their own fear, turned their backs on him. But his sense of responsibility demanded that he be his own person under pressure (see Chapter 2) to help the colleague, who eventually won the battle, even though he later lost the chair position.

However, when politics is seen as a power struggle, Doll becomes uninterested. In our conversations about the issue of power, he mentioned Serres's story about the lamb and the wolf (see Doll, 1993, pp. 32–33, pp. 182–183). Serres reinterpreted La Fontaine's fable of the lamb and the wolf drinking at a stream. While the fable ended with the wolf eating the lamb, Serres warned that it was a dangerous game because one has to *always* be the best in order to get the upper hand, and the shepherd might very well come looking for the lamb and then kill the wolf. Doll brought this story to a Japanese school for a class discussion:

> In my postmodern book, I have that story of Serres about the lamb and the wolf. I told that story to the twelfth-grade students in Japan and posted the question about what it means to be human. The wolf ate the lamb so the wolf has won, but Serres says it means the wolf has lost. How do you interpret the meaning of winner and loser? That generated fascinating discussions. (Interview with Doll, 2010)

This story asks critical questions: What does winning or losing mean? How can we go beyond the game of win-or-lose situations? As an educator, the questions Doll asked Japanese students are important for us in grappling with how to live together better. For Doll, participatory politics is not about a power struggle, but about serving people and sharing life. In this sense, it comes close to ethical relationships.

Pinar (2011, 2013) takes an ethical turn in the U.S. field of curriculum studies as he argues that the primacy of politics has repressed the necessity for ethical self-encounter, relationships with others, and international engagement with the world. The commitment to engage with others and the alterity of the other as an ethical project is shared by Doll and Pinar. Hendry also developed a deep appreciation of Doll's sense of emergent control through team teaching, commenting that "letting go is my politics now" (interview with Hendry, 2012). If letting things emerge in complex class dynamics becomes a form of politics that replaces the power struggle with everyone's participation, then the political becomes the ethical, which shifts the relational dynamic from the win-or-lose mentality to the share-and-transform mentality.

The ethical engagement with the alterity of the other with the potential for the transformation of all participants is open to the uncertain. Giving up control that guarantees certainty is difficult, but without this capacity for resisting imposing knowledge and letting learning occur, the teacher cannot enact ethical relationality in the classroom. For Doll (2013), the possibility of ethics in a postmodern, complex, and global society lies in the "conversation for the Good" in which

> each converser listens attentively to the other—indeed honoring the other's 'otherness.' Each converser speaks from a position of humility, knowing full well that her or his comments may be mistaken. Indeed each teacher in a post-modern frame needs to acquire the art of dealing with the uncertain, not by imposing or dictating authority, but by letting authority be dissipated. Understanding in this frame emerges, is not

dictated, is always provisional, and comes from our wisdom of knowing how to let learning occur. (p. 69)

In a dissipated structure, conversation for the Good can initiate a process of self-organization in which uncertainty and mistakes can play a positive role.

Hendry argues that "[Doll's] political-ness is his spirituality. His attempt to be in the moment with others" while being nonjudgmental and being unconditional was both political and spiritual (interview with Hendry, 2010). The heart of ethical relationality is indeed spiritual for Doll. Both Alcazar and Triche share the Catholic faith with Doll, and spirituality has been an important thread in Doll's pedagogical relationships with them. Both Alcazar and Doll call themselves "heretical Catholics." Alcazar explains the source of this term:

> I told Bill that the Greek word, *hairesis*, really means the other belief that is rejected by the institution. When I was studying theology, I realized sometimes the official statement of an institution is not necessarily the way that the truth should be expressed. . . . So there is, in theological language, the development of truth and the development of the faith. Sometimes when the faith is no longer making sense, it is healthy to keep the other belief, the other teaching, the third instruction, because sometimes the third instruction gives life to the traditional instruction, or sometimes it makes room for the excluded instruction. So in that sense, Bill and I shared the term, heretical Catholic, and I think I can translate loosely in the Greek (*hairesis kata holos*), which means the journey to the wholeness through the other. So if you put the journey to wholeness through the usual path, being heretical means the journey to wholeness through the other. (Interview with Alcazar, 2012)

When an official teaching becomes lifeless, reviving its creative energy requires the infusion of the third, or the other. The journey to wholeness is a process of engaging the third instruction. Alcazar was trained to become a priest, but he realized along the way that even though the Church was committed to charity, it did not really pay much attention to achieving social justice. He was more interested in treating social illness itself rather than its symptoms. Because of his activism on behalf of the disadvantaged in the Philippines, he was tortured and has learned to spiritually journey through the other teaching that is nonconventional. Doll (2002b/2012) has struggled with spirituality together with Alcazar; his journey to wholeness of spirit through the other has been a soul-searching process. It is both an ethical and spiritual journey in which sustained engagement with the other creates a condition for ongoing transformation.

Doll's sense of ethics and spirituality is closely related to his notion of responsibility in multiple dimensions. One is the sense of being committed to others and to the work that one does that he has had since he was a teen searching for his way in society; another is the sense of questioning authority or institutions—the Derridian notion of responsibility that must go beyond the established and be open to difference. He admitted that these notions of responsibility are not necessarily compatible, but the tension within the double gestures is a healthy one. Here he uses the language of science: symbiosis. The coming together of difference in symbiogenesis gives birth to something new over a period of time when differences begin to fuse. "I don't want a new wholeness; I want a new difficulty arising out of that symbiogenesis" to keep creativity alive (interview with Doll, 2010).

Influenced by Steve Mann, who is a radical activist, Doll nevertheless has not followed the path of radical activism, but has chosen to engage students critically and meaningfully in their studies and educational practices in the world. He admits, "I still feel guilty that I was not able to do more to help my students realize the conditions of their society better" (interview with Doll, 2010). Doll is not someone who is unwilling to admit his own limitations. However, at the intersections of politics, ethics, and spirituality bound by a strong sense of responsibility, Doll's pedagogy opens up a complicated space in which each pathway is simultaneously contested and enriched. As much as the dynamics of participation cannot remain static but must be open to uncertainty and incompleteness, the political, the ethical, and the spiritual are all embedded in relationships that emerge from interactions.

William Doll's pedagogy of participation connects with his pedagogy of play, pedagogy of perturbation, pedagogy of presence, pedagogy of patterns, pedagogy of passion, and pedagogy of peace to form a complex web of teaching, life, and thought. His vision—a favorite quote from Milan Kundera—of creating a "fascinating imaginative realm where no one knows the truth and everyone has the right to be understood" (quoted in Doll, 1990/2012, p. 144) is embodied in these 7 P's, which invite other educators' own imagination.

* * * * * * * * * *

A journey marked by complexity and transformation is open-ended, like William Doll's curriculum metaphor of an open heart in the 3 S's. There is no perfect closure for this book as his journey continues, circularly and recursively. I have walked in the labyrinth of Doll's pedagogy with the 7 P's as routes to map out its inner path. These seven themes are not neatly separable but act as recurring threads branching out to form the web of Doll's pedagogy. A labyrinth symbolizes a cycle of life that leads to an ongoing process of rebirth (Shen, 2004). It also indicates "a test or suffering through which to achieve psychic and spiritual growth" (p. 297). Coincidently, one of the classical labyrinths found in many cultures has seven circuits (http://www.labyrinthireland.com/labyrinfo.html). Doll has passed through his tests, and his ongoing journey of transformation is still open-ended.

Going back to the central questions addressed in this book, I think that the relationships between Doll's life experiences and teaching are complex. Doll's first tendency, to be a serious and authoritarian teacher, came from his experiences in school rather than his life experience of a playful childhood. But after he began to teach, play came back to him rather quickly as he learned to respect his students. His teaching experiences also have enriched his life and made him more fully present in life. Particularly his team teaching experiences with women teacher educators have taught him important lessons of enacting dissipative control in making connections with others who are different. Naturally a joyful person, a community organizer, he also has had to learn to craft his pedagogy throughout the curves of his life's path. Both spontaneity and intentional design play a role in his teaching.

The relationship between Doll's teaching in school and his pedagogy in teacher education is not a smooth one either, even though the interplay between the two has been mutually informative. Doll has had to struggle with how to teach pre-service teachers to let go and enable their own students to learn. Teaching in school and teaching about teaching are different and the teacher educator has the dual role of teaching students and teaching about teaching, a role that requires going beyond the teacher educator's modeling because theorizing teaching practices is a necessary step (Loughran, 2006). While Doll does model his approaches in teaching about teaching, which provides a natural bridge between theory and practice for his students, his pedagogical concerns are how students can develop their own teaching approaches that may or may not mirror his own.

The relationship between life and thought is complicated, contradictory, and unpredictable (Wang, 2009). Crucial points in Doll's life such as his playful childhood, the loss of his father, the turbulence of the 1960s, his divorce, and China visits have influenced his philosophical understanding of the world and in turn influenced his teaching, often in implicit ways. However, the mutual influence between life and thought has not been seamless but has remained dynamic and incomplete. The interviews and conversations have given him a deeper understanding of how some of the most influential layers of his life in family and culture have impacted the way he thinks and teaches. As a philosopher and a curriculum scholar, his thought goes beyond pedagogy, but pedagogy has remained one of his central concerns. The healthy tension between his teaching theory and practice throughout his lifetime as an educator has generated creative scholarship and dynamic pedagogy.

In his pedagogy of teacher education, Doll has opened up new possibilities for students to critically reflect on their previous experiences and go beyond the conventional constraints to find alternative approaches. Not following the mainstream, he has walked his own pathways, shedding light on less-explored areas, and encouraged his students to follow their own hearts to less-traveled roads. From the parade child to the king of chaos, the complex journey of William Doll shifts our lens to see the educational landscape anew, through the interactive, joyful, web-making processes of play, perturbation, presence, patterns, passion, peace, and participation. His pedagogy is a gift not only to students but also to all educators.

Am I making Doll into a superhero in this book? That is a question that only readers can answer. I have attempted to portray the complexity of Doll's pedagogical life history in its creative power, including his blind spots and limitations, but his pedagogical and scholarly brilliance can hardly be concealed. Self-organization as his central curriculum metaphor works within and across his life, thought, and pedagogy to bring forth the magic of his teaching. My hope is that this book brings to life a wonderful and inspirational educator—not another conventional superhero—to energize our communal effort to create a better educational world.

Now that we have listened to Doll's stories—ordinary and extraordinary—and the sounds of his footsteps, may we share smiles and laughter with him as we move on to create and walk our own paths?

NOTE

1. The Bergamo Conference on Curriculum Theory and Classroom Practice, an important conference in the U.S. field of curriculum studies, is associated with the Reconceptualization movement initiated in the late 1970s (Pinar, 2000). Started as the site of engaging alternative intellectual conversations that were excluded by the mainstream curriculum discourses, the Bergamo conference, sponsored by the *Journal of Curriculum Theorizing*, has creatively shaped the landscape of curriculum studies as we know it today. For details about the history of the Bergamo conference, see Janet Miller's (2005) and William Pinar's (1999, 2008) recountings.

BILL DOLL'S PEDAGOGY

A Peek Behind the Curtain

David Kirshner

For twenty years Bill Doll was my colleague at Louisiana State University, perhaps partly mentor, but more friend and supporter than guide. Although I share in the curriculum theory community that Bill nurtured, and co-taught some courses with him, my primary allegiance has been to the mainstream pedagogical enterprise from which curriculum theory recoils. Still, my allegiance is conditioned by a sense of the alienation that scientistic application of psychology to teaching has provoked in the work of my friend Bill and many other colleagues at LSU and beyond; my goal embraces opening up the standard paradigm so it can be seen to be not quite so antagonistic to the humanist impulse that (fortunately) grounds the educational enterprise for many in the academy.

I intend here to render Bill Doll's pedagogy using a framework that reinterprets the relation of psychology to educational practice, while remaining rooted in psychology's scientific insights and accomplishments. The goal is multifaceted: to honor an inspired teacher and leader and probe his multifaceted talents; to demonstrate how the varied paradigms of psychology can be marshaled to embrace widely divergent pedagogical aspirations, touching on, also, those espoused within curriculum theory; and to address, a little, the tensions within education that come from a separation of humanist and scientific enterprises.

My story of Bill Doll's pedagogy is one that unfolds in context within context within context. The outermost context is the history of pedagogical debate over the past century. Bill has been mindful of that history and frames his own efforts in reference to it. The middle context is that of the Department of Curriculum & Instruction that, during the twenty-year Bill Pinar / Bill Doll era, participated in the broader history in a particular way. The innermost context is that of my relationship with Bill Doll realized within, and as part of, the dynamic of the department. One thing I've learned from Bill Pinar and Bill Doll is that 'personal' matters. I honor that teaching by starting with this innermost context.

I was hired into LSU's C&I department in fall 1987, a semester or two before Bill Doll and a year or so following Bill Pinar's arrival as chair. Bill Pinar had come with the intention, and the mandate from Dean Billy Smith, to put LSU on the map as a center for the *reconceptualist* movement of curriculum theory. Resisting the technocratic tendency that has colonized institutionalized education, reconceptualists seek to provide a space in educational practice for the creative and generative human impulse to find expression. Reconceptualists draw on standard narratives of oppression such as critical theory, but also on a wide range of intellectual traditions—history, politics, psychoanalysis, phenomenology, postmodernism, autobiography, aesthetics, theology—thereby replicating, fractal-like, the freedom sought within schooling through the theorizing of schooling meant to provoke that freedom. Thus reconceptualism is twice marginal; from the mainstream ideology of technocratic efficiency that governs institutionalized education, and from the standard traditions of critique that seek a fixed counter narrative.

Perhaps it is my own sense of marginal connection to the mainstream that made me an attractive candidate to fill the mathematics education position at LSU—if not an ally, at least not a purveyor of the hubris of the center. Regardless of the reason, the curriculum theory program welcomed me to its social network, and has remained respectfully receptive to my intellectual contributions. For my part, the inclusion has been much appreciated, not only because the energy of curriculum theorists is contagious and exciting, but also because intellectual strands in my own work were initiated through the curriculum theory connection. For instance, my reader on situated cognition theory, co-edited with then–LSU colleague Tony Whitson (Kirshner & Whitson,

1997), originated in my reading of Valerie Walkerdine's post-structural analysis of mathematics learning introduced to me by curriculum theory graduate student Linda Dugas in a course she took with me.

The spirit of the program throughout the twenty-year participation of the Bills has been one of a fecund intercourse of the personal, the social, and the intellectual. The praxis of the curriculum theory group encompassed all of these aspects in a disciplined and integrated way, including mentoring of students and junior faculty, hosting of parties and conferences, co-editing and co-authoring of books and papers, and of course through creation and maintenance of university courses—all of this accomplished within a framework of peripheral participation always receptive to the aspirations of participants for a more central role in the program. It continues to be a privilege to be affiliated with this fecund program.

Over the years, a large cadre of faculty came to LSU as core members of the curriculum theory program, later institutionalized as the Curriculum Theory Project, including Bill Stanley, Tony Whitson, Denise Egea-Kuehne, Cameron McCarthy, Leslie Roman, John Davies, Jacques Daigneault, Wendy Kohli, Mary-Ellen Jacobs, Nina Asher, Claudia Eppert, Kaustuv Roy, and Jayne Fleener, as well as Petra Hendry, Jackie Bach, and Roland Mitchell who continue the program today. Collectively, this group has mentored dozens—perhaps hundreds—of master's students and doctoral researchers, many of whom have gone on to establish distinguished careers themselves as curriculum theorists. Importantly, the LSU community of curriculum theory has included the full complement of core faculty, graduate students, and affiliate faculty who came, to a lesser or greater extent, to care about the problematic of education as a human project within the strictures of technocratic institutions.

Some graduate students, already committed to the study of curriculum, came to LSU attracted by the reputation of our core faculty. Many more were nurtured into the program through their participation in courses and their inclusion in the peculiar blend of personal, social, and intellectual engagements that is the dynamic constitution of curriculum theory at LSU. To a large extent it became Bill Doll who established and maintained the pathways for novice curriculum theorists to progress into full membership. He did this through courses that awoke students to their potential as valued co-contributors to curriculum theory; through Friday Friends, a moveable feast/discussion group of current students, faculty, and alumni convening in a local restaurant following his Friday morning course; through after-school cocktails at local bars organized by his students; and through Curriculum Camp, an annual retreat/graduate student conference that serves as a springboard to professional presentation and publication. At any one time, Bill maintained dozens of personal relationships with current and former students, critiquing their work, encouraging their progress, calming their insecurities, inviting further participation, applauding successes, and shining as a beacon of support and stability through what might be a time of difficult personal transition for them. Of course, Bill Doll was not alone in these efforts. All core faculty co-participated in maintaining the community of curriculum theory at LSU. But it was Bill Doll who was the touchstone for entry into the program for students beginning their exploration of curriculum theory.

Bill Doll's Vision

So, how did Bill do it—especially his courses? How did he organize instruction in such a way as to provide transformative experiences for so many students? To understand his pedagogy, it is important to be acquainted with the scope of ideas that informed, yes, the content of his instruction, but also the methods. Bill Doll's early scholarship in education and philosophy was oriented by the pragmatist efforts of James, Dewey, Mead, and others to theorize a functional psychology of learning that took as its unit of analysis not the mechanism of learning abstracted from life lived, but the fully engaged individual pursuing life projects, interacting in a social world, and rooted in spiritual quest. An engaged pedagogy was the goal, one that involved students in mutual community, in connection with the world outside the school, and that built up genuine enactive competence rather than decontextualized skills. Yet a student of history, not just philosophy, Bill understood all too well that the Progressive Education Movement embodying these ideals had stood not a chance against the juggernaut of technocratic education rooted in a "scientific" psychology that demanded as its price the dissociation of learning from life lived (Lagemann, 1989). The straight rows of children chained to their desks became emblematic for Bill of imprisoned spirits, the doors to Education now ruthlessly guarded by the snarling, yapping dogs of science and mathematics, straining against their leashes to maul and maim any so foolish as to venture an alternative vision.

Bill's sense of personal intimidation by science and mathematics should not be underestimated, neither the brilliance of his strategy to tame these beasts through postmodern reinterpretation. Roots of postmodernism lie in impasses to the modernist vision that emerged within mathematics and science; the paradoxes of logic, the quantum impossibilities that come to characterize the material substrate. Arising, in part, from these anomalies is the postmodern playfulness of shifting frames of reference, narrative structure, and wordplay. Not that the dogs of science and mathematics suddenly jump into your lap and lick your face. Rather, this is a long-term mission building toward a more hopeful future; a focus on creation of new pedagogical possibilities in the here and now, as the earthquake of postmodernism slowly reassembles the grounding of our collective cultural consciousness. As I perceive it, this is the vision that Bill Doll nourishes.

Chaos and complexity theory came to occupy a special place for Bill in the artifacts of postmodernism that frame a new educational possibility. As mathematician Abraham (2000) characterized it, "The Chaos Revolution is a bifurcation event in the history of the sciences, comprised of sequential paradigm shifts in the various sciences. Perhaps it is also a major transformation in the world of cultural history" (p. 78). Arising from the mathematics of recursivity, chaos/complexity charts a non-deterministic, yet patterned, pathway; a semi-stable oasis of regularity suspended between deterministic certainty and utter breakdown into hyper-kinetic incoherence. During the period between the dissolution of his marriage of twenty-five years and his eventual remarriage, Bill confided to me that his own life was following the chaotic path of a *strange attractor*; riding this curve enabling him to steer between the certainty of depression and the dissociation of utter fantasy. One way or another, I believe riding the curve serves for Bill as the metaphor for a completely engaged instruction. The question is, what does this mean for instructional practice?

Bill Doll's Chaotic Pedagogy

The dynamic of chaotic pedagogy plays out in Bill's instruction as a surfeit of emergent possibility overflowing the manageable. What this means in practice is that an activity is established—say, theoretical discussion of texts or a mathematical exploration—that has a certain structure. Certain resources are provided, certain generative possibilities for combination highlighted. As classroom engagement progresses, the constrained practices of the activity feed back on themselves, leading to an acceleration of new possibilities. The result is a sense, at the interactive level, of being mutually overwhelmed by possibilities, coordinated at the individual level with awe at the generative capacity of one's own system of ideation. Of course, this doesn't always work so well, but when it does work it forms the basis for personal empowerment embedded in community. In Bill's expansive pedagogy it works often enough to sustain and grow a dynamic community of scholarship.

Here are a couple of examples from my participation in Bill Doll's classes. The first harks back to a series of five weekly summer math sessions we conducted with a group of fifth graders, perhaps in the early 1990s. I recall one session organized by Bill that explored the meaning of the fraction as a ratio of numerator to denominator. The details are murky, but a fraction like 1/2 was the subject of scrutiny. Various versions and non-versions of 1/2 were explored so that what was compared was the multiplicative ratio of the numerator to the denominator across cases. For instance, 8/24 involves a tripling of the numerator to get the denominator whereas 3/6 involves only a doubling. So in multiplicative terms the numerator 8 is puny with respect to its denominator, 24, in comparison to 3 in relation to its denominator, 6. Needless to say, thinking about fractions in this way was a novelty for students typically taught nothing more in mathematics classes than procedural routines for completion of exercises.

As I recall, the class dynamic involved consideration of many fraction instances offered by the students, in each case comparing the multiplicative ratio of numerator to denominator. Somehow the examples built on one another until at some point the numerator for consideration in a given fraction was itself a fraction. The students were perplexed by this eventuality, but it was obvious in looking at him that Bill was entirely in his element; the exhortation to the group to think it through, to apply to this strange newcomer what previously had been working so well, to persist and extend meaning so that in a few minutes whatever ideas students may have had about fractions had become blown open to incorporate an entirely new category of thing—the fraction-fraction. Just lovely!

The second example presents not a classroom episode, but a pedagogical method I observed while sitting in on one of Bill's graduate seminars, maybe in 2004 or 2005—a method of *disjoint readings* I believe he employed frequently. The tool, simply, is assignment of different readings to different groups of students within the class; perhaps a choice of readings. What is striking about this method, apart from its somewhat haphazard juxtaposition of conceptual elements—I'll return to that later—is the possibilities it creates for dynamic interplay of conversational contributions as one student responds to another from a completely different frame. I argue that this kind of kinetic conversation is the primary means Bill uses to break out beyond boundaries, to *overwhelm* and *inspire*—his twin learning vehicles so well-captured in his trademark "Whooooosh!"

Learning, Learning, Learning

In this section, I introduce a framework for analysis of pedagogy rooted in psychology, but defiant of many of its presumptions (Kirshner, 2016), and use it to probe the essence and efficacy of Bill Doll's teaching. My sociology of science viewpoint picks up on Kuhn's (1970) characterization of science's "preparadigm" phase (p. 17) in which the multiple paradigms of a young science compete existentially with one another to establish an initial consensus across the field that unites the science and signals the achievement of paradigmatic maturity. Crucially, the competing paradigms are "incommensurable" (Kuhn, 1970, p. 4)—theorists "see different things, and they see them in different relations to one another" (p. 150). Thus persuasion of the merits of one's paradigm must depart from the logical and rational argumentation we associate with mature sciences. Typically, the paradigms of psychology compete in hegemonic fashion by claiming to already (or immanently) capture the essential aspects of other paradigms (e.g., Skinner's [1958] attempt to extend behaviorism from unmediated response conditioning to verbal behavior famously rebuffed by Chomsky [1959] at the start of the "cognitive era"; and cognitive psychology's attempt to project the successes of decontextualized puzzle solving to contextual reasoning, leading to establishment of situated cognition theory [Brown, Collins, & Duguid, 1989; Lave, 1988]).

The ironic consequence of this competitive dynamic is a universal tendency for learning theorists to declare learning a unitary or multifaceted construct (of course, as captured in the theorist's own paradigm!). Although, this is clearly a counterfactual representation of learning theory, it prevails in education owing to psychology's historic domination (Lagemann, 2000).

The Genres Approach

Apprehending psychology as the multi-paradigmatic science that it is, rather than as the uni-paradigmatic science it hopes to become, enables us to observe a deep synergy between the learning goals of education and the varied notions of learning that motivate psychologists. For as Fletcher (1995) noted, our culture's "folk psychology is built into scientific psychological theories in a more thoroughgoing fashion than is commonly realized" (p. 97). In particular, the National Council for Accreditation of Teacher Education's (2002) model of valued learning goals—skills, knowledge (concepts), and dispositions (cultural practices)—tracks rather well with major branches of psychology—behavioral (Skinner), developmental (Piaget), and sociocultural (Vygotsky), respectively. Although, as we shall see, the fit is not quite exact, characterization of good teaching in *genres*, each informed by an independent theorization of learning, enables a clarity and focus for pedagogical theory not possible in our usual discursive frames that seek to apprehend learning as a unitary construct.

It is beyond the scope of this chapter to provide more than a cursory sketch of the full genres framework (Kirshner, 2016); I focus here most fully on the theorizations and teaching methods that bear upon the analysis of Bill Doll's pedagogy.

Teaching of skills (routine responses to pre-set problem types) is informed by behavioral psychology as well as by the Implicit Learning program within cognitive psychology (Reber, 1993). Learning, in this interpretation, is development of subcognitive correlations among stimulus and response elements. This incremental learning process is supported by repetitive exposure to the regularities of a task domain through performance of routine exercises.

Teaching of concepts is informed by Piaget's developmental psychology, specifically, the complementary processes—dynamically driven by disequilibrium—of *assimilation* of elements of the environment to existing conceptual structures, and *accommodation* of cognitive structures to better fit with elements of the environment (Piaget, 1977a). Crucially, the process of conceptual restructuring depends on experiencing tensions and discrepancies in one's existing conceptual structures (Piaget, 1977b). To facilitate this learning process, the teacher, based on a model (always tentative) of the student's current conceptual structures, devises a task environment intended to juxtapose discrepant aspects of current understanding. However, simply provoking cognitive conflict is not sufficient. The teacher also envisions "the flow of the students' (changing) thinking" (Simon et al., 2010, p. 91) along a "hypothetical learning trajectory" (Clements & Sarama, 2004; Simon, 1995), and seeks to modulate the learning experience accordingly.

The learning of cultural practices through participation in a cultural milieu is most directly studied in sociocultural psychology initiated by Vygotsky.

However, sociocultural psychology is complicated by the desire to account for "the dual process of shaping and being shaped through culture" (Cole, 1996, p. 103). Instead, I turn for theoretical grounding to sociologist Talcott Parsons's (1951) notion of unconscious "introjection or internalization of the [cultural] standard so that to act in conformity with it becomes a need disposition in the actor's own personality structure" (p. 37).

What constitutes an appropriate pedagogical method for this form of learning hinges on whether or not the students are culturally identified with the reference culture. If so, the teacher's role consists largely of modeling and coaching; students, seeking to acculturate themselves to the reference culture, emulate the cultural practices modeled by the teacher. This *acculturation pedagogy* can be contrasted with the *enculturation pedagogy* appropriate for students who do not aspire to membership in the reference culture. This pedagogical practice involves nurturing the evolution of desired cultural practices so that they come to be "interactively constituted by each classroom community" (Yackel & Cobb, 1996, p. 475). In this way, students learn through introjection of cultural norms, rather than from the teacher, in a direct sense. Importantly, this teaching agenda must remain tacit. Requiring students to adopt markers of a culture with which they are not identified may create "intrapersonal conflict" for the student (Brown, 2004, p. 810).

Bill Doll's Pedagogy

The notions of learning addressed earlier reference our culture's folk psychology reflected also in the paradigms of psychology (Fletcher, 1995). In this respect, the genres approach is offered not as an ad hoc set of pedagogical methods, but as a schematization of the basic intuitions that drive pedagogical practice across the broad educational enterprise.

Evidence for this assertion is marshaled in Kirshner (2016) through the method of *crossdisciplinary analysis* in which pedagogical methods and associated educational issues are refracted through the trifocal lens of the genres approach. Performing a crossdisciplinary analysis is a matter of evaluating whether skills are promoted, whether concepts are promoted, and whether dispositions are promoted. If it seems the answer to any of these is yes, it is incumbent on the analyst to identify the particular skills, concepts, or dispositions being addressed, to evaluate the efficacy of the pedagogy with reference to the indicated genre, and to assess whether the methodology employed is appropriate with respect to the characteristics of the student body. Employing this method has produced surprising insights into a broad range of educational issues, including: Skinner's programmed instruction; the Reading Wars and the Math Wars; metacognition; reform pedagogy, critical pedagogy, identity, and ethical issues (Kirshner, 2016).

Bill Doll's pedagogy falls under the broad umbrella of reform methods in which "knowledge is personally constructed and socially mediated" (Windschitl, 2002, p. 137). Reform methods tend to downplay skills, falling along a continuum that encompasses concepts and dispositions to varying degrees (Alexander, 2007). The case I want to make here is that Bill's pedagogy lies at the dispositions extreme of the reform spectrum, eschewing support for students' learning of particular concepts in favor of induction of students into valued cultural practices. I should underscore that crossdisciplinarity is non-evaluative with respect to the choice of learning goals, the emphasis being on the particular character and efficacy of the methods.

The observation that Bill's pedagogy is oriented toward support of dispositions only, and not concepts, might seem unlikely, even counterintuitive, given the challenging philosophical texts in which he regularly immerses his students. Indeed, Bill does intend that students learn to struggle with concepts—that they become *conceptual thinkers*. But this is a dispositional goal, a form of cultural participation, quite distinct from staking out particular conceptual difficulties he wants his students to encounter and work through. Nor does he seek to orchestrate specific cognitive conflicts as is needed to promote conceptual restructuring. For instance, his method of "disjoint readings" discussed earlier (assigning different readings to different students) pushes in exactly the opposite direction, creating a dynamic of community stress, rather than the intramental stress associated with cognitive conflicts at the individual level that progress along a previsioned hypothetical learning trajectory (Simon et al., 2010). Of course, students in his classes *do* work through specific concepts, and develop stronger understandings as a result. But this is "advertent learning" (Kirshner, 2016, p. 135)—learning that may be expected to arise from instruction, but that is not directly supported within the structure of the pedagogy.

Turning to cultural practices, we can highlight academic/philosophical culture as the reference culture into which Bill seeks to inculcate his students. The pedagogy is acculturationist. The cultural agendas are explicit, and the community of the classroom is extended to be co-extensive with the authentic academic culture of the university. Students' papers can be refined into publication submissions, and Bill's own in-progress work frequently finds its way into the curriculum. Thus, his own practices of thinking and writing become available as cultural models for his students.

His pedagogy also employs enculturationist methods. Within that classroom culture, practices of academic/philosophical culture that he manifests become increasingly evident: the drawn out, reflective struggle over ideas; the savoring of academic language; the appreciation of lateral/metaphorical thinking characteristic of "complex conversations"; the playing with ideas. Over the course of a semester, one observes these dispositional practices becoming increasingly normative within the classroom community. Thus students learn by emulation and by immersion.

The other dynamic unfolding within the context of Bill's classes is identity transformation. Although nominally, in the U.S. tradition, all students are equal, in Bill's practice there is a hierarchy of peripheral participation from novice students to more central participants. As a regular feature of his instruction, students sit in a circle, which indicates co-equal opportunity to participate. And all students—novice and advanced—have occasional responsibility to report on readings and lead discussions. Although rarely (perhaps, never) adjudicating the "correctness" of their ideas, Bill consistently and honestly celebrates the sophistication of students' writing and thinking. Thus, Bill offers students a way forward toward increasingly central participation in the community through creation of more sophisticated scholarly writing. Many students accept this invitation.

An example of Bill's enculturationist influence is offered by Marie Capoccia (personal communication, June 26, 2011), one of his former students:

> I believe one of the most prominent characteristics of Dr. Doll was the reaction he had to an interesting/thought provoking statement. I think that anyone who participated in his class would remember this vividly. Leaning back in his chair, he would close his eyes and smile. This was one of my favorite experiences in his class—it was a visual example of what it means to truly enjoy learning.

Crucially, Bill's classroom is interpenetrated with the actual academic community through social events that include faculty and students; through lectures of visiting scholars; through Curriculum Camp, which offers a chance to develop and test-drive academic papers; through participation and presentation at the annual Bergamo conference and now the AAACS conference; and through co-publication with faculty. These make the classroom a seamless entryway into actual academic membership. Note that the moral critique associated with acculturationist pedagogies for students not already culturally identified with the reference culture does not arise in this case, as the identity invitation is optional. Students are not differentially graded based on their manifestation of target dispositions.

The Reference Culture

Although academic/philosophical culture has been identified as the reference culture, this is an imprecise designation. For instance, an analytic philosopher would likely create an acculturationist pedagogy incorporating hard-edged judgments of the rigor of students' logical inferences. Or a classical philosopher would evaluate whether or not the student has mastered the received interpretations of canonical philosophical thought. Such pedagogies would admirably prepare students to enter into the culture of the respective philosophical traditions, but are remote from the culture of scholarship inculcated in Bill's classroom. What, then, is the particular variety of academic/philosophical culture that Bill Doll strives to preserve and promote?

What I want to suggest is that this inclusive intellectual community bonded together in personal relationship and humanist values, in which creative impulse trumps the strictures of intellectual orthodoxy, is a utopian culture of educational scholarship presaging a broader utopian society, namely, the society that eventually emerges in an educational system ordered by those same values. This is a vision of societal transformation through education that accords well with John Dewey's aspirations for an enlightened school culture as "the deepest and best guarantee of a larger society which is worthy, lovely, and harmonious" (Dewey, 1900, p. 44). That the core experience of this educational culture, as realized in Bill's classroom, is a sense of awe within community suggests religious inspiration. Postmodernism in robes, I give you Bill Doll, a father of curriculum theory.

REFERENCES

Abraham, R. (2000). The chaos revolution: A personal view. In R. Abraham & Y. Ueda (Eds.), *The chaos avant-garde: Memories of the early days of chaos theory* (pp. 81–90). Singapore: World Scientific.

Alexander, P. A. (2007). Bridging cognition and socioculturalism within conceptual change research: Unnecessary foray or unachievable feat? *Educational Psychologist, 42*(1), 67–73. Retrieved from www.tandfonline.com/doi/pdf/10.1080/00461520709336919

Brown, B. A. (2004). Discursive identity: Assimilation into the culture of science and its implications for minority students. *Journal of Research in Science Teaching, 41*(8), 810–834.

Brown, J. S., Collins, A., & Duguid, P. (1989). Situated cognition and the culture of learning. *Educational Researcher, 18*(1), 32–42.

Chomsky, N. (1959). A review of B. F. Skinner's *Verbal behavior*. *Language, 35*(1), 26–58.

Clements, D. H., & Sarama, J. (Eds.). (2004). Hypothetical learning trajectories [Special issue]. *Mathematical Thinking and Learning, 6*(2).

Cole, M. (1996). *Cultural psychology*. Cambridge, MA: The Belknap Press.

Dewey, J. (1900). *School and society*. Chicago, IL: University of Chicago Press.

Fletcher, G. (1995). *The scientific credibility of folk psychology*. Mahwah, NJ: Lawrence Erlbaum.

Kirshner, D. (2016). Configuring learning theory to support teaching. In L. English & D. Kirshner (Eds.), *Handbook of international research in mathematics education* (3rd ed., pp. 98–149). New York: Taylor & Francis.

Kirshner, D., & Whitson, J. A. (Eds.). (1997). *Situated cognition: Social, semiotic, and psychological perspectives*. Mahwah, NJ: Lawrence Erlbaum.

Kuhn, T. S. (1970). *The structure of scientific revolutions*. Chicago, IL: University of Chicago Press.

Lagemann, E. C. (1989). The plural worlds of educational research. *History of Education Quarterly, 29*(2), 185–214. doi:10.2307/368309

———. (2000). *An elusive science: The troubling history of education research.* Chicago, IL: University of Chicago Press.

Lave, J. (1988). *Cognition in practice.* Cambridge, England: Cambridge University Press.

National Council for Accreditation of Teacher Education. (2002). *Professional standards for the accreditation of schools, colleges, and departments of education.* Washington, DC: Author.

Parsons, T. (1951). *The social system.* London, England: Routledge & Kegan Paul.

Piaget, J. (1977a). Problems of equilibration. In M. H. Appel & L. S. Goldberg (Eds.), *Topics in cognitive development* (Vol. 1, pp. 3–14). New York: Plenum.

———. (1977b). *Recherches sur l'abstraction réfléchissante* (Vols. 1 & 2). Paris, France: Presses Universitaires de France.

Reber, A. S. (1993). *Implicit learning and tacit knowledge: An essay on the cognitive unconscious* (Oxford Psychology Series No. 19). Oxford, England: Oxford University Press; New York: Clarendon Press.

Simon, M. A. (1995). Reconstructing mathematics pedagogy from a constructivist perspective. *Journal for Research in Mathematics Education, 26*(2), 114–145.

Simon, M. A., Saldanha, L., McClintock, E., Karagoz Akar, G., Watanabe, T., & Ozgur Zembat, I. (2010). A developing approach to studying students' learning through their mathematical activity. *Cognition and Instruction, 28,* 70–112.

Skinner, B. F. (1958). *Verbal behavior.* New York: Macmillan.

Windschitl, M. (2002). Framing constructivism in practice as the negotiation of dilemmas: An analysis of the conceptual, pedagogical, cultural, and political challenges facing teachers. *Review of Educational Research, 72*(2), 131–175. Retrieved from http://rer.sagepub.com/content/72/2/131.full.pdf+html

Yackel, E., & Cobb, P. (1996). Sociomathematical norms, argumentation, and autonomy in mathematics. *Journal for Research in Mathematics Education, 27*(4), 458–477.

AFTERWORD

Of Experiencing Pedagogy by Rainbow Light

Molly Quinn

In the Kabbalah, it is said that human beings are Divine Sparks, the glowing emanations of God's creative light, each one unique and irreplaceable. However, some sparks burn brighter than others. Bill Doll is one of those rare individuals whose Being burns so bright that his mere presence transforms the lives of all those he touches. He has transformed my life.

I first met Bill in the fall of 1990. God had "died" (Nietzsche 1883/1982), my religious life work had become questionable—me, in slight existential crisis. I enrolled in a class with Dr. Doll to get recertified as a teacher. Whoosh![1] And the rest is history—a PhD, and paradigmatic shift—an utterly transformed cosmology and consciousness later for me! This towering yet remarkably inviting man stood before me, a veritable postmodern pastiche; a combination of Bostonian gentleman, Northeastern Intellectual, Mystic Philosopher-Sage, Bohemian Artist-Poet, Dr. Seuss, and Santa Claus: exuberant, big hearted (albeit thin), playful, poetic—dare I say, eccentric; yet elegant, distinguished, and wise. Gorgeous snowy white hair, the classic bow tie, and a laugh that could change the world; he taught me via Piaget (1977) to appreciate my "disequilibrium" (Doll, 1989/2012c),[2] to embrace chaos (1986/2012h) as a cosmological good, to honor my inner complexity (1998/2012i), and "just Dewey it" (1990/2012g) with respect to the writing. As a teacher, he accomplished what he aspired to create: the "art born as the echo of God's laughter," "a fascinating imaginative realm" "where no one owns the truth and everyone has the right to be understood" (Kundera, 1988, cited in Doll, 2006/2012e, p. 231).

Life, and we could add education, and curriculum, like New York City—where I now live and work, is a "beautiful catastrophe" (Le Corbusier, 1935). Bill helped me see and experience more of its beauty. I learned there was creative, educational fruit in the quest itself, in asking and seeking, in the struggle for meaning (Doll, 2008/2012d). I learned that in this postmodern world, we may celebrate and bear witness to the unpresentable, the inexplicable, the mystery and uncertainty (Doll, 2002/2012j). I received from Bill, as he so eloquently puts it: "an education which questions the being of all we hold sacred while at the same time manifests a faith that such questioning will lead us to the sacredness of being" (2002/2012j, p. 42). My God may have died (and that was a good thing), but I certainly have known divine grace through Bill Doll's presence in my life.

"Every blade of grass has its Angel that bends over it and whispers: 'Grow! Grow!'" says the Talmud (as cited in Lurie, 2013, p. 95). Bill, thank you for being my Angel, and noble Patriarch—the beloved guardian of so many of us here—for being my Mentor, my Guide, my Advocate, my Colleague, my Friend. I love you, "Dad"—my intellectual father. Or is it "Dollface," according to your undergraduate students? You are a real doll, indeed, you know. I must agree with Emerson[3]—especially in your case—that: "What lies behind us, and what lies before us are tiny matters compared to what lies within us." Of course, what lies behind you are great richness and accomplishment we are here to honor. And what lies before you are great adventure and blessing we are here to toast. But mostly, we are here to celebrate glorious, glowing you, for, post-structuralism notwithstanding, what lies within you is beyond language—exceeding, surpassing, words. Thus, we come together to bask in your radiant, transcending light, and thank the gods above for our dear Divine Doll.

Happy Birthday, King of Chaos, King of Our Hearts! All my love, always.

(Quinn, 2001, February)

It has been fifteen years since I offered this tribute at the In Praise of the Post-Modern conference in celebration of the seventieth birthday of my beloved teacher, William E. Doll, Jr. Just recently, I spoke with him on the phone as he enjoyed a day of festivities at his home in Victoria, British Columbia, Canada, on the occasion of his eighty-fifth birthday. Time has passed and much has changed. Bill is no longer at my alma mater, Louisiana State University, where I studied with him as a doctoral student in my hometown, and I no longer live in New York City. And while the two of us talk and email from time to time, it is not as much as I should like, and I see him even more rarely. In fact, as I listened to him speak last May in Ottawa at the International Association for the Advancement of Curriculum Studies Conference (IAACS); inspired, enlightened, challenged, uplifted, as always by his words and ideas, I simultaneously felt a sense of yearning and loss, realizing just how much I missed my teacher, his daily living presence in my life and world. I also felt an immense sense of gratitude to have had and have and know such a teacher and presence in my life. What wisdom, humility, eloquence, imagination, and love! In this, much has not changed at all.

What has not changed, too, are these generative gifts and blessings he has given me, which continue to work in and on me, dynamic, educative, transforming. As his student, his influence, I trust, lives on in this way in my own life, scholarship, service, and especially in my own teaching. My prayer is that I bring even a tiny spark of such divine light to my students, as well, to also carry forward. In concert with this hope, I have much joy and delight in this work of Hongyu Wang on Doll's pedagogy and work as a teacher educator, and in relation to his life history and intellectual thought—particularly for the opportunity it affords others to learn much from Bill—for so many of us to continue to learn, to learn anew, from him—too. As Doll himself has expressed elsewhere in acknowledging fruitful engagements with and gifts from others: "The academic genealogy is a joy to behold. There is comfort in knowing that one's ideas will sprout elsewhere" (2012a, p. xviii). While I was not among the many interviewed—students, former students, colleagues, and family members—here toward crafting this rich portrait from Doll's teaching palette, my experiences with and of Bill resonate and harmonize much with what their stories unveil, and it is a great honor to me to participate in this project, and bring to the work my own "story-sharing" (Wang, this volume) regarding the treasures we might glean in the study of Doll's life and pedagogy. This offering has already been a fresh spur to my own further growth and renewal, by which I have been reminded: "Development is not continual and gradual; it is punctuated with plateaus, spurts, and bifurcation points" (Doll, 1986/2012h, p. 142); from which we may perpetually hope—a new opening, a new conversation, a new understanding, is always possible, for ourselves as well as for our students. Doll dwells in such hope, and fosters it in others.

Because of Hongyu's generosity, I was able to use the unpublished manuscript as a text in a doctoral course I co-taught this semester on advanced theories of learning and pedagogy. We coupled it with Truiet's (2012) collection of Doll's essays for selecting the person of Bill Doll—his pedagogical theories and practices—as a kind of case study to be examined in concert with various frames on learning theory and pedagogy. Wang aptly points out that there has been little research in teacher education respecting the life experiences of teacher educators and their pedagogical theories and practices, or in relation to their intellectual explorations. Too often attention is directed to the methodological and technical dimensions of teaching, which belies so much of its richness, complexity, and humanness. Too often pedagogy remains bent on best practices and evidence-based strategies amid meager attempts to acknowledge context and culture while holding on to control, prediction, and even imposition, and often as construed through the lens of positivist, modernist science. Doll, of course, critiques such, challenges our givens, and shows us a different way, and in a different world and universe—the one opened up via the new sciences of chaos and complexity, even as he engages myriad other intellectual traditions as well (Doll, 2008/2012d, p. 22).

Despite and perhaps because of Doll's lack of directed interest in subjectivity—his embrace of a kind of dispersed, dissipative subjectivity, dynamically and systemically situated—the educational adventure Wang explores through him illumines the profound power of the kind of pedagogical presence he embodies, and pedagogical postures he adopts, in his person, and practice. Trueit (2012) notes that "Doll's intellectual trajectory through pragmatism, post-modernism, and complexity theory testifies to his intellectual work, but his intimate relation to the terrain of educational thought makes this journey much more of a pilgrimage" (p. xii). In studying such, embracing the relational, situational, and emergent, we are emboldened to—as Doll would call it—*recurricularize* the pedagogies and pilgrimages of our own lives, contemplate the ways in which what we live and experience, what we study, might be implicated in and experimented with via our own teaching to advance transformational growth in ourselves and in those we teach, and to foster intellectual and even spiritual community.

"We live in a rainbow of chaos." In the essay "Looking Forward," Doll (2006/2012e, p. 228) showcases these words of Paul Cézanne. The 7 P's of Doll's pedagogy presented by Wang seem to me to represent an effort to study and understand the brilliance of Doll, this king of chaos, and his extraordinary pedagogy, "magical ability to teach," by rainbow light, via examination of its varied colors which shine all the more brightly when juxtaposed and yet are also seen whole, by a "holistic matrixical methodology" (this volume). And more mystic than mathematician, I think too of the number 7 in sacred geometry (Schneider, 1995)—a vehicle of life for Pythagoras, a union of tripartite spirit and matter as outlined in four elements; the seven virtues, stages of man [sic], steps of Buddha, chakras, Islamic heavens and earths, Hindu worlds, branches of the Jewish menorah, harmonic notes, liberal arts: divine spark, emanation of God's creative light. The rainbow is also a sign of blessing and promise. As a kid I called the crystal a rainbow maker, casting a kaleidoscope of color upon the window or wall at the touch of sunlight, a dream maker. Indeed, Bill's pedagogy has blessed me, and visited me time and again as promise and possibility, and also as spur to the making of my own dreams as an educator. I suspect this spectral window into it here has in some manner prompted such similarly for readers. As Wang so poignantly recapitulates via her 7 P's:

> From the parade child to the king of chaos, the complex journey of William Doll shifts our lens to see the educational landscape anew, through the interactive, joyful, web-making process of play, perturbation, presence, patterns, passion, peace, and participation. His pedagogy is a gift not only to students but also to all educators.

For me, it was a gift indeed, and one wholly unexpected. It was quite by accident that I found myself in that first graduate course, Elementary Curriculum, with Bill at LSU (although I doubt there are, in truth, such things as "accidents"). Anticipating a rather unenlightening, straight-forward education class in the study of curriculum, I was met with this incredibly exuberant man, a powerful and yet reassuring presence: immeasurably and infectiously enthusiastic about the work in which he was engaged, it felt as if he was warmly inviting us as students "to explore [the] subject's intricacies through the act of play" (Doll, 1999/2012l, p. 208), to play with him, together, with these intriguing questions and thoughts he brought before us, and encouraged us to bring before the class, as well. Believing that the "vitality of knowledge and life cannot be maintained without playing with ideas" (Wang, this volume), with Whitehead (1929/1967), he taught us to "throw ideas into every combination possible" (p. 2). He was so effective in rousing us to such play, in my first paper I toyed with a children's fairy tale, beginning it thus: *"Fe, fi, fo, fum, I smell the stench of the curriculum."* My mother worried I wouldn't pass my first graduate course.

We read a good number of diverse and difficult texts that semester, and wrestled with all kinds of questions and content. Yet, Bill read and wrestled with us; as we wrote and shared commentaries in response to course readings and class discussions, Bill engaged in the same, modeling a kind of critical and creative inquiry I had never experienced before. Listening deeply, he reflected our questions back to us and affirmed them, while ever challenging us to grow beyond them, to journey further, to explore deeper, bigger, questions. Of this approach, Doll (2009/2012b) states:

> Asking students to articulate what they see, hear, and read, and to listen to the interpretations of others opens a dialogue ... that provides an opportunity for the *new* to emerge. If we ask where the *new* comes from, the answer lies in interpretive inquiry; inquiry where we and the "other," no matter what or who the "other" may be (person, idea, fact, culture) come into interactive play. (p. 22)

While I had been exposed to the work of Jean Piaget innumerable times before in my undergraduate education classes, in Bill's class I was encouraged to read Piaget himself for the first time, and through Bill's presentation of his ideas was able to grasp them in a totally new light—they were given a larger context, and thus meaning. I will never forget Bill's chalkboard drawing of an amoeba-like organism moving from a state of equilibrium to disequilibrium in the direction of a higher equilibration, and his use of this evolutionary process to discuss the importance of confusion and even chaos in our own learning, as well as in that of our students. Experiencing such perturbation in my own life at the time, this fruitful focus, even on failure and waste as educative, was heartening to me. One classmate created a lively cartoon depicting a student who said, "I'm totally lost. I can't keep on living"; to which she had Doll respond, "Yes. Wonderful! Wonderful! Keep going!" Here, "In the context of teaching, perturbations, errors and confusions that are traditionally dismissed as negative elements become key elements for initiating students' intellectual reorganization to a higher level" (Wang, this volume)—tapping into "the spirit of creativity hidden in every situation, yet to be born" (Doll, 2003/2012f, p. 104).

I remember, too, him exhorting us in our reading of and response to the curriculum historian Herbert Kliebard (2004) to have something of a "love affair" with the author, his ideas, or a "confrontation," a "knock-down drag-out argument" if need be—typical for Bill, but it had never occurred to me before nor had I been taught to engage with any text so fully, so avidly. "To know a situation we need to 'plunge into the situation' [Dewey, 1934/1958], struggle with it, explore it, in depth, *be* in it, intuiting, feeling our way around," Doll (2008/2012d, p. 31) would say. I was presented with John Dewey's (1922/1964) "What Is Freedom?" in this class and was personally moved by it in relation to my life situation. Bill then directed me to reflect through it on my own education and the ways in which it had served to both thwart and foster my growth and liberation. Inciting passion, inspiring presence—and exemplifying both, he nudged me in the direction of uncovering the existential patterns within my own biographical situation from which might emerge new meanings, and through which I might find peace.

"In Doll's pedagogy of presence, relationality lies at the heart of teaching and learning. He provides pedagogical companionship and sustained engagement to contribute to students' own sense-making and meaning-making" (Wang, this volume). He also promoted participation, creating a supportive intellectual community that allowed us all to openly share our ideas and feelings with others in class. He respected us enough, and our intelligence, to even present scholarly manuscripts on which he was working and with which he was struggling, welcoming criticism and advice—embracing "Dewey's notion of the teacher as a *prima inter pares* (first among equals)" (Doll, 2009/2012b, p. 17). He advocated and embodied Richard Rorty's dialogic imperative to "keep the conversation going" (1979, p. 377)—his pedagogical creed itself grounded in a reflective relationship between teacher and student that enlists the student to join with the teacher in inquiry (Doll, 1993, p. 160).

The study of curriculum, and education, through his synthesizing brilliance and pioneering vision took on profound meaning through which I was changed—it involved a breadth and depth of inquiry that was tied to my own well-being, my own existential questions demanding attention, my own personal path and calling in life, even as it was also inextricably connected to all the world, to cultivating our very humanity in harmony with the life of the cosmos via education. I studied the history of curriculum, and was exposed to different educational theories and pedagogical practices, but I also learned "the examined life"—to read, reflect, question, think, in ways that opened up new horizons for me, and never-before-dreamed-of possibilities. Additionally, I was introduced to the fascinating worlds of chaos theory, cosmology, fractal geometry, hermeneutics, postmodernism, and process philosophy—to name but a few domains of thought Bill artfully drew on to elucidate educational understanding. Here was a teacher, indeed, who adored study, the life of inquiry, and guiding others in their discovery of its treasures and transformative possibilities. And he has continued in this, his gift and his passion—inspiring and having inspired me, and a host of others across the globe.

After that, I had the privilege of taking many other, equally enticing courses with Bill, and could describe them in almost as equal detail because of their impact on my person and thought. In fact, Bill—ever attuned to the web of relationships and "pattern which connects" (2002/2012j, p. 36)—gleaned something of my life's calling when I could not, recruited me for doctoral study, and served as my major advisor—shepherding me through the challenging and exciting journey to the PhD and beyond as a mentor throughout my pilgrimage as a professor. I could go on with much "story-sharing" and many colorful stories, and yet here have tried to organically illumine just something of the way these 7 P's of his pedagogy have been experienced by and affected me as a student, and in but only my first course of study with him, now many years ago. The reader might think me unrealistically laudatory. Certainly, Bill and I have not always agreed, and I have played my fair share of the part of the prodigal daughter too. But, undoubtedly, Bill has been among my greatest of teachers. He embodies all the qualities and dispositions of an extraordinary teacher in his very person: With unparalleled generosity of heart and expansiveness of mind, he is ever, it seems, sharing his wisdom and faith and passion for and joy in life—and for inquiring into it—with others.

Both in and out of the classroom, he inspires students to find the same within themselves, and thereby be transformed, and to become energized to approach their own work with this exemplary integrity, vitality, and vision. There are few whose life and pedagogy could teach us as much. It is a privilege to be called his student, to be able to call him my teacher, and to speak if but briefly of his pedagogical art here.

> The teacher's task is not to impart all he, or she, or society knows, but to help the learner grow in the development of his or her own powers of thought. This is a very real educational need to have the learner re-invent the wheel. (Doll, 1981/2012k, p. 195)

Doll's wheel is a magnificent—if not spinning, fractaled—color wheel: bright, brilliant, sparkling rainbow light. May we soak up its sunshine, and its reflecting, refracting luminescent rays—the 7 P's of his pedagogy presented herein. And may such implore us to the reinvention of our own wheels too—our own pedagogical potentials and possibilities. Paint, palette, divine pedagogical Doll light; are we ready to play?

NOTES

1. I am referencing an expression Doll himself often utters in response to some new connection, insight, or idea emerging in conversation.
2. While I did not originally cite sources for this live, orally presented tribute, I have added them here for the reader's reference. For this purpose, I have used the collection of Doll's essays (Trueit, 2012), published well after the date of this occasion, and directed the reader, in some cases, also to essays originally written after the time period in which I studied with him as well as after this toast was crafted. Of course, the influence of Doll as my teacher and mentor, and my study of his work, continued after my doctoral studies and has endured up to the present day.
3. This quote is most often attributed to Emerson, though others like Henry David Thoreau and Oliver Wendell Holmes, Jr. have also been identified as its author as well. Quote investigator (http://quoteinvestigator.com/2011/01/11/what-lies-within/) locates its first use in an anonymously published book entitled *Meditations in Wall Street* (1940), New York: William Morrow. I credit Emerson here, with my source for the citation: J. Cameron, with M. Bryan (1992), *The Artist's Way* (p. 6). New York: Putnam.

REFERENCES

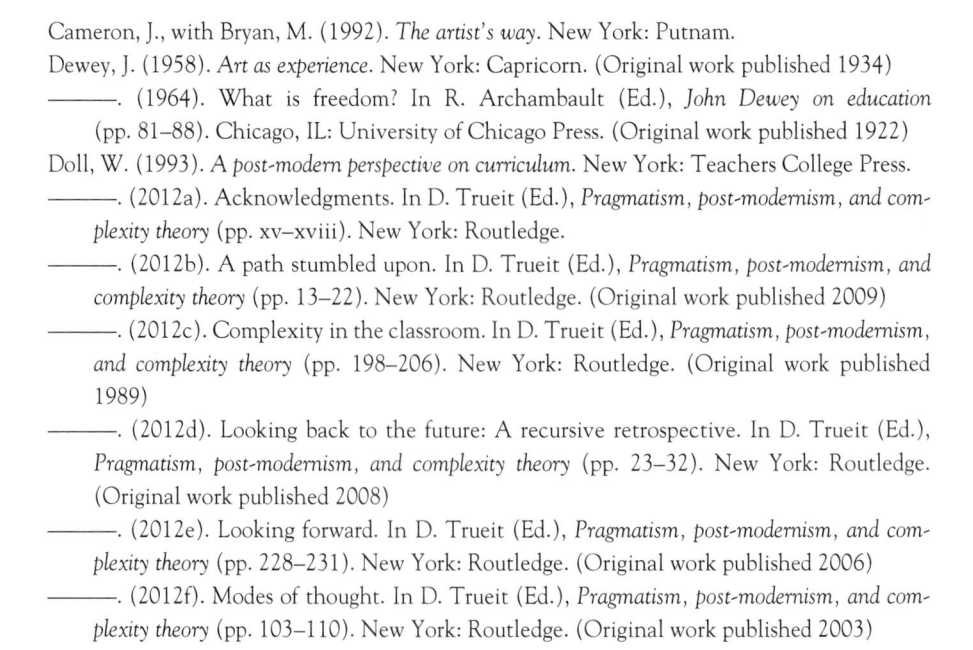

Cameron, J., with Bryan, M. (1992). *The artist's way*. New York: Putnam.
Dewey, J. (1958). *Art as experience*. New York: Capricorn. (Original work published 1934)
———. (1964). What is freedom? In R. Archambault (Ed.), *John Dewey on education* (pp. 81–88). Chicago, IL: University of Chicago Press. (Original work published 1922)
Doll, W. (1993). *A post-modern perspective on curriculum*. New York: Teachers College Press.
———. (2012a). Acknowledgments. In D. Trueit (Ed.), *Pragmatism, post-modernism, and complexity theory* (pp. xv–xviii). New York: Routledge.
———. (2012b). A path stumbled upon. In D. Trueit (Ed.), *Pragmatism, post-modernism, and complexity theory* (pp. 13–22). New York: Routledge. (Original work published 2009)
———. (2012c). Complexity in the classroom. In D. Trueit (Ed.), *Pragmatism, post-modernism, and complexity theory* (pp. 198–206). New York: Routledge. (Original work published 1989)
———. (2012d). Looking back to the future: A recursive retrospective. In D. Trueit (Ed.), *Pragmatism, post-modernism, and complexity theory* (pp. 23–32). New York: Routledge. (Original work published 2008)
———. (2012e). Looking forward. In D. Trueit (Ed.), *Pragmatism, post-modernism, and complexity theory* (pp. 228–231). New York: Routledge. (Original work published 2006)
———. (2012f). Modes of thought. In D. Trueit (Ed.), *Pragmatism, post-modernism, and complexity theory* (pp. 103–110). New York: Routledge. (Original work published 2003)

———. (2012g). Post-modernism's utopian vision. In D. Trueit (Ed.), *Pragmatism, post-modernism, and complexity theory* (pp. 144–152). New York: Routledge. (Original work published 1990)

———. (2012h). Prigogine: A new sense of order. In D. Trueit (Ed.), *Pragmatism, post-modernism, and complexity theory* (pp. 134–143). New York: Routledge. (Original work published 1986)

———. (2012i). Recursions on complexity. In D. Trueit (Ed.), *Pragmatism, post-modernism, and complexity theory* (pp. 163–168). New York: Routledge. (Original work published 1998)

———. (2012j). Struggles with sprituality. In D. Trueit (Ed.), *Pragmatism, post-modernism, and complexity theory* (pp. 33–42). New York: Routledge. (Original work published 2002)

———. (2012k). The educational need to re-invent the wheel. In D. Trueit (Ed.), *Pragmatism, post-modernism, and complexity theory* (pp. 193–197). New York: Routledge. (Original work published 1981)

———. (2012l). Reflections on teaching. In D. Trueit (Ed.), *Pragmatism, post-modernism, and complexity theory* (pp. 207–221). New York: Routledge. (Original work published 1999)

Kliebard, H. M. (2004). *The struggle for the American curriculum* (3rd ed.). New York: Routledge.

Kundera, M. (1988). *The art of the novel* (L. Asher, Trans.). New York: Grove Press.

Le Corbusier, C.-E. J. (1935). Battery City Park inscription, New York City. In V. Luiselli, December 9, 2014. Collected poems. *The New Yorker*.

Lurie, A. (2013). Faith and spirituality in the workplace: A Jewish perspective. In *Handbook of faith and spirituality in the workplace* (pp. 85–101). New York: Springer.

Nietzsche, F. (1982). Thus spake Zarathustra. In W. Kaufmann (Ed. & Trans.), *The portable Nietzsche* (pp. 102–440). New York: Penguin. (Original work published 1883)

Piaget, J. (1977). *The essential Piaget* (H. Gruber & J. Voneche, Eds.). New York: Basic Books.

Rorty, R. (1979). *Philosophy and the mirror of nature*. Princeton, NJ: Princeton University Press.

Schneider, M. (1995). *A beginner's guide to constructing the universe: The mathematical archetypes of nature, art and science—a voyage from 1 to 10*. New York: Harper Perennial.

Trueit, D. (Ed.). (2012). *Pragmatism, post-modernism, and complexity theory*. New York: Routledge.

Whitehead, A. N. (1967). *The aims of education and other essays*. New York: The Free Press. (Original work published 1929)

REFERENCES

Abbott, A. Edwin. (1992). *Flatland*. New York: Dover. (Original work published 1884)
Addams, Jane. (2007). *Newer ideals of peace*. Champaign: University of Illinois Press. (Original work published 1906).
Aoki, Ted T. (1990). Inspiriting the curriculum. In Ted T. Aoki (2005), *Curriculum in a new key* (William F. Pinar & Rita L. Irwin, Eds.) (pp. 357–365). Mahwah, NJ: Lawrence Erlbaum.
———. (1992). Layered voices of teaching. In Ted T. Aoki (2005), *Curriculum in a new key* (William F. Pinar & Rita L. Irwin, Eds.) (pp. 187–197). Mahwah, NJ: Lawrence Erlbaum.
———. (2005). *Curriculum in a new key* (William F. Pinar & Rita L. Irwin, Eds.). Mahwah, NJ: Lawrence Erlbaum.
Arends, Robert L., Masla, John A., & Weber, Wilford A. (1971). *Handbook for the development of instructional modules in competency-based teacher education programs*. Syracuse, NY: Center for the Study of Teaching.
Art Matrix. (1990). *Mandelbrot and Julia sets* [Video]. (Available from Art Matrix, PO Box 880, Ithaca, NY 14851–0880)
Ayers, William. (2002). Creating the teacher and changing the world. In Elijah Mirochnik & Debora C. Sherman (Eds.), *Passion and pedagogy* (pp. 37–51). New York: Peter Lang.
Ball, Deborah Loewenberg, & Cohen, David K. (1999). Developing practice, developing practitioners. In Linda Darling-Hammond & G. Sykes (Eds.), *Teaching as the learning profession* (pp. 3–32). San Francisco, CA: Jossey-Bass.
Bateson, Gregory. (1972). *Steps to an ecology of mind*. New York: Ballantine.

———. (1988). *Mind and nature*. Toronto, Canada: Bantam Books. (Original work published 1979)

Biesta, Gerta. (2004). "Mind the gap" communication and the educational relation. In Charles Bingham & Alexander M. Sidorkin (Eds.), *No education without relation* (pp. 11–22). New York: Peter Lang.

Bingham, Charles. (2004). Let's treat authority relationally. In Charles Bingham & Alexander M. Sidorkin (Eds.), *No education without relation* (pp. 23–37). New York: Peter Lang.

Boler, Megan. (1999). *Feeling power*. New York: Routledge.

Bowles, Douglas F. (1973). Competency-based teacher education? The Houston story. *Educational Leadership, 30*, 510–512.

Briggs, John. (1992). *Fractal*. New York: Touchstone.

Britzman, Deborah P. (2003). *Practice makes practice*. Albany, NY: SUNY Press. (Original work published 1991)

Broom, Jeffrey W. (2011). Investigating relationships: Thoughts on the pitfalls and directions. *Complexity, 1*, 38–43.

Burns, Donna M. (2010). *When kids are grieving*. Thousand Oaks, CA: Corwin.

Capra, Fritjof. (1996). *The web of life*. New York: Anchor.

Chen, Yuting. (2014). From follower to creator. In William F. Pinar (Ed.), *Curriculum studies in China* (pp. 69–82). New York: Palgrave Macmillan.

Chodakowski, Anne, Egan, Kieran, Judson, Gillian C., & Stewart, Kym. (2010). Some neglected components of teacher education program. In Cheryle J. Craig & Louise F. Deretchin (Eds.), *Cultivating curious and creative minds* (Part 2, pp. 5–21). Lanham, MD: Rowman & Littlefield Education.

Cloninger, Kevin, & Mengert, Christina. (2010). In pursuit of joy. In Cheryle J. Craig & Louise F. Deretchin (Eds.), *Cultivating curious and creative minds* (Part 1, pp. 4–23). Lanham, MD: Rowman & Littlefield Education.

Cochran-Smith, Marilyn, & Zeichner, Kenneth M. (Eds.). (2005). *Studying teacher education*. Mahwah, NJ: Lawrence Erlbaum.

Cochran-Smith, Marilyn, Feiman-Nemser, Sharon, & McIntyre, D. John. (Eds.). (2008). *Handbook of research on teacher education* (3rd ed.). New York: Routledge.

Crowell, Sam, & Reid-Marr, David. (2010). The non-linear nature of emergent teaching. *Complicity, 7*(2), 116–119.

Darling-Hammond, Linda, & Bransford, John. (2005). *Preparing teachers for a changing world*. San Francisco, CA: Jossey-Bass.

Davis, Brent, & Samara, Dennis. (2002). Constructivist discourses and the field of education. *Educational Theory, 52*(4), 409–428.

Derrida, Jacques. (1987). Letter to a Japanese friend. In Jacques Derrida (1991), *A Derrida reader* (Peggy Kumuf, Ed.) (pp. 269–276). New York: Columbia University Press.

———. (1990). Privilege. In Jacques Derrida (2002), *Who is afraid of philosophy?* (Jan Plug, Trans.). Stanford, CA: Stanford University Press.

———. (1992). *The other heading* (Pascale-Anne Brault & Michael B. Naas, Trans.). Bloomington: Indiana University Press.

Dewey, John. (1997). *Democracy and education*. New York: The Free Press. (Original work published 1916)

Doll, Mary Aswell. (1995). *To the lighthouse and back*. New York: Peter Lang.

Doll, William E., Jr. (1977). The role of contrast in the development of competence. In Alex Molnar & J. Zahorik (Eds.), *Curriculum theory* (pp. 50–63). Washington, DC: Association for Supervision and Curriculum Development.

———. (1984). Developing competence. In William E. Doll, Jr. (2012), *Pragmatism, post-modernism, and complexity theory* (Donna Trueit, Ed.) (pp. 66–80). New York: Routledge.

———. (1986). Prigogine. In William E. Doll, Jr. (2012), *Pragmatism, post-modernism, and complexity theory* (Donna Trueit, Ed.) (pp. 134–143). New York: Routledge.

———. (1989). Complexity in the classroom. In William E. Doll, Jr. (2012), *Pragmatism, post-modernism, and complexity theory* (Donna Trueit, Ed.) (pp. 198–206). New York: Routledge.

———. (1990). Post-modernism's utopian vision. In William E. Doll, Jr. (2012), *Pragmatism, post-modernism, and complexity theory* (Donna Trueit, Ed.) (pp. 143–152). New York: Routledge.

———. (1993). *A post-modern perspective on curriculum*. New York: Teachers College Press.

———. (1998a). The spirit of curriculum. *Early Childhood Education, 31*, 3–7.

———. (1998b). Curriculum and concepts of control. In William F. Pinar (Ed.), *Curriculum* (pp. 295–323). New York: Garland.

———. (1999). Reflections on teaching. In William E. Doll, Jr. (2012), *Pragmatism, post-modernism, and complexity theory* (Donna Trueit, Ed.) (pp. 207–221). New York: Routledge.

———. (2000). Classroom management. In William E. Doll, Jr. (2012), *Pragmatism, post-modernism, and complexity theory* (Donna Trueit, Ed.) (pp. 222–227). New York: Routledge.

———. (2002a). Beyond methods? Teaching as an aesthetic and spirit-ful quest. In Elijah Mirochnik & Debora C. Sherman (Eds.), *Passion and pedagogy* (pp. 127–152). New York: Peter Lang.

———. (2002b). Struggles with spirituality. In William E. Doll, Jr. (2012), *Pragmatism, post-modernism, and complexity theory* (Donna Trueit, Ed.) (pp. 33–42). New York: Routledge.

———. (2002c). Ghosts and the curriculum. In William E. Doll, Jr., & Noel Gough (Eds.), *Curriculum vision* (pp. 23–70). New York: Peter Lang.

———. (2002d). Beyond methods. In William E. Doll, Jr. (2012), *Pragmatism, post-modernism, and complexity theory* (Donna Trueit, Ed.) (pp. 81–97). New York: Routledge.

———. (2003). Modes of thought. In William E. Doll, Jr. (2012), *Pragmatism, post-modernism, and complexity theory* (Donna Trueit, Ed.) (pp. 103–110). New York: Routledge.

———. (2004). Crafting an experience. In William E. Doll, Jr. (2012), *Pragmatism, post-modernism, and complexity theory* (Donna Trueit, Ed.) (pp. 98–99). New York: Routledge.

———. (2005). Keeping knowledge alive. In William E. Doll, Jr. (2012), *Pragmatism, post-modernism, and complexity theory* (Donna Trueit, Ed.) (pp. 111–119). New York: Routledge.

———. (2008). Looking back to the future. In William E. Doll, Jr. (2012), *Pragmatism, post-modernism, and complexity theory* (Donna Trueit, Ed.) (pp. 23–32). New York: Routledge.

———. (2009). A path stumbled upon. In William E. Doll, Jr. (2012), *Pragmatism, post-modernism, and complexity theory* (Donna Trueit, Ed.) (pp. 13–22). New York: Routledge.

———. (2011a). Enlarging the space of the possible. *Complicity*, 8(1), 32–37.

———. (2011b, October). *Our mistake is in our thinking.* Keynote speech at the 32nd Bergamo Conference on Curriculum Theory and Classroom Practice, Dayton, OH.

———. (2011c). Da Xia lecture. In William E. Doll, Jr. (2012), *Pragmatism, post-modernism, and complexity theory* (Donna Trueit, Ed.) (pp. 232–243). New York: Routledge.

———. (2011d). Structures of the postmodern. In William E. Doll, Jr. (2012), *Pragmatism, post-modernism, and complexity theory* (Donna Trueit, Ed.) (pp. 131–160). New York: Routledge.

———. (2012). *Pragmatism, post-modernism, and complexity theory* (Donna Trueit, Ed.). New York: Routledge.

———. (2013). An exploration of "ethics" in a post-modern, complex, global society. *Transnational Curriculum Inquiry*, 10(2), 64–70.

Doll, William E., Jr., Fleener, Jayne, M., Trueit, Donna, & St. Julien, John. (Eds.). (2005). *Chaos, complexity, curriculum, and culture.* New York: Peter Lang.

Doll, William E., Jr., & Gough, Noel (Eds.). *Curriculum visions.* New York: Peter Lang.

Doll, William E., Jr., & Trueit, Donna. (2010). Thinking complexly. In William E. Doll, Jr. (2012), *Pragmatism, post-modernism, and complexity theory* (Donna Trueit, Ed.) (pp. 172–188). New York: Routledge.

Edelfelt, Roy A., & Raths, James D. (1999). *A brief history of standards in teacher education.* Reston, VA: Association of Teacher Educators.

Edgerton, Susan Huddleson. (1991). Particularities of "otherness." In Joe L. Kincheloe & William F. Pinar (Eds.), *Curriculum as social psychoanalysis* (pp. 77–98). Albany, NY: SUNY Press.

Farber, Jerry. (2008). Teaching and presence. *Pedagogy*, 8(2), 215–225.

Fleener, Jayne (2002). *Curriculum dynamics.* New York: Peter Lang.

Fowler, Leah C. (2005). *A curriculum of difficulty.* New York: Peter Lang.

Fraser, James W. (2007). *Preparing America's teachers.* New York: Teachers College Press.

Garrison, Jim. (1997). *Dewey and eros.* New York: Teachers College Press.

Gleick, James. (1987). *Chaos.* New York: Penguin Books.

Goodson, Ivor F. (Ed.). (1992). *Studying teachers' lives.* London, England: Routledge.

Goodson, Ivor F., & Sikes, Pat. (2001). *Life history research in educational settings.* Buckingham, England: Open University Press.

Hansen, David T. (2011). *The teacher and the world.* London, England: Routledge.

Hendry, Petra Munro. (2012). *Engendering curriculum history.* New York: Routledge.

Hershock, Peter D. (2012). *Valuing diversity.* Albany, NY: SUNY Press.

Holmes Group. (1986). *Tomorrow's teachers.* East Lansing, MI: Author.

hooks, bell. (2000). *Feminism is for everybody.* Cambridge, MA: South End Press.

Houston, W. Robert. (1990). *Handbook of research on teacher education.* New York: Macmillan.

Huebner, Dwayne. (1999). *The lure of the transcendent.* (Vikki Hillis, Ed.; William F. Pinar, Intro.). Mahwah, NJ: Lawrence Erlbaum.

———. (2000). Curriculum as concern for man's temporality. In William F. Pinar (Ed.), *Curriculum studies* (pp. 237–249). Troy, NY: Educator's International Press. (Original work published 1975)

Jardine, David, Friesen, Sharon, & Clifford, Patricia. (2012). *Curriculum of abundance*. Mahwah, NJ: Lawrence Erlbaum.

Jung, Carl. (1969). *The archetypes and the collective unconscious* (2nd ed.) (R. F. C. Hull, Trans.). Princeton, NJ: Princeton University Press.

Kane, Ruth. (2007). From naïve practitioner to teacher educator and researcher. In Tom Russell & John Loughran (Eds.), *Enacting a pedagogy of teacher education* (pp. 60–76). London, England: Routledge.

Kauffman, Stuart. (1995). *At home in the universe*. New York: Oxford University Press.

Kliebard, Herbert M. (2004). *The struggle for the American curriculum, 1893–1958* (3rd ed.). New York: RoutledgeFalmer.

Kluth, Paula, & Straut, Diana. (2003). Do as we say and as we do. *Journal of Teacher Education*, 54(3), 228–240.

Korthagen, Fred, Loughran, John, & Russell, Tom. (2006). Developing fundamental principles for teacher education programs and practices. *Teaching and Teacher Education*, 22, 1020–1041.

Kosnik, Clare. (2007). Still the same yet different. In Tom Russell & John Loughran (Eds.), *Enacting a pedagogy of teacher education* (pp. 16–30). London, England: Routledge.

Kristeva, Julia. (2002). *Intimate revolt* (Jeanine Herman, Trans.). New York: Columbia University Press.

Kroll, Linda R. (2007). Constructing and reconstructing the concepts of development and learning. In Tom Russell & John Loughran (Eds.), *Enacting a pedagogy of teacher education* (pp. 95–105). London, England: Routledge.

Labaree, David F. (2004). *The trouble with Ed schools*. New Haven, CT: Yale University Press.

———. (2008). An uneasy relationship. In Marilyn Cochran-Smith, Sharon Feiman-Nemser, & D. John McIntyre (Eds.), *Handbook of research on teacher education* (3rd ed., pp. 290–305). Washington, DC: Association of Teacher Educators.

Latta, Margaret Macintyre. (2013). *Curriculum conversations*. New York: Routledge.

Letterman, Margaret R., & Dugan, Kimberly B. (2004). Team teaching a cross-disciplinary honors course. *College Teaching*, 52(2), 76–79.

Levine, Arthur. (2006). *Educating school teachers*. Washington, DC: The Education Schools Project.

Li, Xin. (2002). *The Tao of life stories*. New York: Peter Lang.

Lin, Jing, Brantmeier, Edward J., & Bruhn, Christa. (2008). *Transforming education for peace*. Charlotte, NC: Information Age.

Loughran, John. (2006). *Developing a pedagogy of teacher education*. London, England: Routledge.

———. (2013). Pedagogy. *Curriculum Inquiry*, 43(1), 118–141.

Maheux, Jean-François, & Lajoie, Caroline. (2010). On improvisation in teaching and teacher education. *Complicity*, 8(2), 86–92.

Marsh, Colin J., & Willis, George. (2007). *Curriculum* (4th ed.). Upper Saddle River, NJ: Pearson.
Martusewicz, Rebecca R. (2001). *Seeking passage*. New York: The New Press.
Mayes, Clifford. (2005). *Jung and education*. Lanham, MD: Rowman & Littlefield Education.
———. (2010). *The archetypal hero's journey in teaching and learning*. Madison, WI: Atwood.
McGuire, Michael. (1991). *An eye for fractals*. Redwood City, CA: Addison-Wesley.
Miller, James. (1993). *The passion of Michel Foucault*. New York: Simon & Schuster.
Miller, Janet L. (1990). *Creating spaces and finding voices*. Albany, NY: SUNY Press.
———. (2005). *Sounds of silence breaking*. New York: Peter Lang.
———. (2010). Communities without consensus. In Erik Malewski (Ed.), *Curriculum studies handbook* (pp. 95–100). New York: Routledge.
Mirochnik, Elijah, & Sherman, Debora C. (2002). *Passion and pedagogy*. New York: Peter Lang.
Morson, Gary Saul, & Emerson, Caryl. (1990). *Mikhail Bakhtin: Creation of a prosaic*. Stanford, CA: Stanford University Press.
Munro, Petra. (1998). *Subject to fiction*. Buckingham, England: Open University Press.
Nevin, Ann I., Thousand, Jacqueline S., & Villa, Richard A. (2009). Collaborative teaching for teacher educators—what does the research say? *Teaching and Teacher Education*, 25(4), 569–574.
Oxford, Rebecca. (2014). *Understanding peace cultures*. Charlotte, NC: Information Age.
Palmer, Parker J. (2003). Teaching with heart and soul: Reflections on spirituality in teacher education. *Journal of Teacher Education*, 54(5), 376–385.
Pinar, William F. (1999). *Contemporary curriculum discourses*. New York: Peter Lang.
———. (2000). *Curriculum studies: The reconceptualization*. Troy, NY: Educator's International Press.
———. (2003). *International handbook of curriculum research*. Mahwah, NJ: Lawrence Erlbaum.
———. (2006). *The synoptic text today and other essays*. New York: Peter Lang.
———. (2008). JCT today. *Journal of Curriculum Theorizing*, 24(1), 4–10.
———. (2009). *The worldliness of a cosmopolitan education*. New York: Routledge.
———. (2011). *The character of curriculum studies*. New York: Palgrave Macmillan.
———. (2013). *Curriculum studies in the United States*. New York: Palgrave Macmillan.
———. (2014). *Curriculum studies in China*. New York: Palgrave Macmillan.
———. (2015). *Educational experience as lived: Knowledge, history, alterity*. New York: Routledge.
Pinar, William F., Reynolds, William, Slattery, Patrick, & Taubman, Peter. (1995). *Understanding curriculum*. New York: Peter Lang.
Porche-Frilot, Donna. (2002). Perspective on Wang. In William E. Doll, Jr., & Noel Gough (Eds.), *Curriculum visions* (pp. 300–303). New York: Peter Lang.
Pratt, Sarah Smitherman. (2011). Emerging changes in teacher education. *Complicity*, 8(1), 43–49.
Prigogine, Ilya. (1983, May). Interview. *Omni*, 84–92; 120–121.
Prigogine, Ilya, & Stengers, Isabelle. (1984). *Order out of chaos*. New York: Bantam.
Quinn, Molly. (2001). *Going out, not knowing whither*. New York: Peter Lang.
———. (2014). *Peace and pedagogy*. New York: Peter Lang.

Reynolds, Sherrie. (2005). Patterns that connect. In William E. Doll, Jr., Jayne M. Fleener, Donna Trueit, & John St. Julien (Eds.), *Chaos, complexity, curriculum, and culture* (pp. 263–276). New York: Peter Lang.
Robertson, Emily. (2008). Teacher education in a democratic society. In Marilyn Cochran-Smith, Sharon Feiman-Nemser, & D. John McIntyre (Eds.), *Handbook of research on teacher education* (3rd ed., pp. 27–44). New York: Routledge.
Romano, Rosalie M. (2002). A pedagogy that supposes passion. In Elijah Mirochnik & Debora C. Sherman (Eds.), *Passion and pedagogy* (pp. 365–377). New York: Peter Lang.
Rorty, Richard. (1989). *Contingency, irony and solidarity.* Cambridge, England: Cambridge University Press.
Russell, Tom, & Loughran, John. (Eds.). (2007). *Enacting a pedagogy of teacher education* (pp. 16–30). London, England: Routledge.
Schubert, William H., & Ayers, William C. (Eds.). (1991). *Teacher lore.* New York: Longman.
Serres, Michel. (1997). *The troubadour of knowledge* (Sheila Faria Glaser & William Paulson, Trans.). Ann Arbor: University of Michigan Press. (Original work published 1991)
Shen, Heyong. (2004). *Analytic psychology.* Beijing: Life, Reader, and New Knowledge Press.
Sikula, John P. (1996). *Handbook of research on teacher education.* New York: Macmillan Reference Books.
Slattery, Laura, Butigan, Ken, Pelicaric, Veronica, & Preston-Pile, Ken. (2005). *Engage.* Oakland, CA: Pace e Bene Press.
Stengel, Barbara S. (2004). Knowing is a response-able relation. In Charles Bingham & Alexander M. Sidorkin (Eds.), *No education without relation* (pp. 139–152). New York: Peter Lang.
Taubman, Peter. (1992). Achieving the right distance. In William F. Pinar & William M. Reynolds (Eds.), *Understanding curriculum and phenomenological and deconstructed text* (pp. 216–233). New York: Teachers College Press.
Thompson, Clarissa S. (2006). Powerful pedagogy. *Teaching and Teacher Education, 22*(2), 194–204.
Todd, Sharon. (2003). *Learning from the other.* Albany, NY: SUNY Press.
Tröhler, Daniel. (2011). *Languages of education.* New York: Routledge.
Trueit, Donna. (2002). Speaking of ghosts . . . In William E. Doll, Jr., & Noel Gough (Eds.), *Curriculum visions* (pp. 269–280). New York: Peter Lang.
Trueit, Donna, Doll, William E., Jr., Wang, Hongyu, & Pinar, William. (2000). *The internationalization of curriculum studies.* New York: Peter Lang.
Walter, Katya. (1994). *Tao of chaos.* Shaftesbury, England: Element.
Wang, Hongyu. (2004). *The call from the stranger on a journey home.* New York: Peter Lang.
———. (2005). Chinese aesthetics, fractals, and the *Tao* of curriculum. In William E. Doll, Jr., Jayne M. Fleener, Donna Trueit, & John St. Julien (Eds.), *Chaos, complexity, curriculum, and culture* (pp. 299–314). New York: Peter Lang.
———. (2008). "Red eyes." *Multicultural Perspectives, 10*(1), 10–16.
———. (2009). Life history and cross-cultural thought. *Transnational Curriculum Inquiry, 6*(2), 37–50.

——. (2010). Intimate revolt and third possibilities. In Erik Malewski (Ed.), *Curriculum studies reader* (pp. 374–386). New York: Routledge.

——. (2014). *Nonviolence and education*. New York: Routledge.

Wang, Hongyu, & Hoyt, Wu Mei. (2007). Sounds of silence breaking. *Journal of the American Association for the Advancement of Curriculum Studies, 3*.

Wang, Hongyu, & Olson, Nadine. (2009). *A journey to unlearn and learn in multicultural education*. New York: Peter Lang.

Wasko-Flood, Sandra. (2010). Labyrinths for creativity and peace in schools. In Cheryle J. Craig & Louise F. Deretchin (Eds.), *Cultivating curious and creative minds* (Part 2, pp. 144–159). Lanham, MD: Rowman & Littlefield Education.

Westheimer, Joel. (2008). Learning among colleagues. In Marilyn Cochran-Smith, Sharon Feiman-Nemser, & D. John McIntyre (Eds.), *Handbook of research on teacher education* (3rd ed., pp. 756–784). New York: Routledge.

Wheatley, Margaret J. (1989). *Leadership and the new science*. San Francisco, CA: Berrett-Koehler.

Whitehead, A. N. (1967). *The aims of education and other essays*. New York: The Free Press. (Original work published 1929)

INDEX

A

Abbott, A. Edwin, 179
Abundance, 7
Acculturationist teaching, 162
Activism, 130, 148, 149
Addams, Jane, xxii, 111, 116, 136, 179
Alcazar, Al, xxviii, xvii, xxxii, 45, 46, 53, 92–94, 100, 122, 123, 130, 138, 145, 146, 148
Alienation, 153
Alterity, 39, 147
Aoki, Ted T., ix–xii, xx, xxii, 3, 21, 45, 82, 135
Asher, Nina, 155
Awe, xiv, 69, 71, 129, 164
Ayers, William, xxvii, 55, 91, 96

B

Bach, Jackie, 155
Bakhtin, Mikhail, 122
Bateson, Gregory, 22, 39, 40, 71, 73, 88, 96, 121
Becoming, 34–35
Behaviorism, 36, 65, 141, 159
Bell, Daniel, 98, 141
Bergamo Conference, xxviii, 142, 152, 163
Berry, Thomas, 22
Biesta, Gerta, 50
Bildung, 12, 14
Bingham, Charles, 50
Body, 13, 51, 82, 95, 116
Boler, Megan, 30
Borhani, Maya Tracy, xxxii, 12
Both/and, xii, 28
Bransford, John, 17
Brantmeier, Edward J., 116
Briggs, John, 101

Britzman, Deborah P., 95
Bruhn, Christa, 116
Broom, Jeffrey W., 86
Brunner, Jerome, 6, 7, 32, 63, 74, 75, 121, 145

C

Canada, 33, 46, 82, 83, 89, 102, 103, 106, 146, 168
Capra, Fritjof, 23, 84
Catholic faith, xv, 22, 148
Change, xvii, 2, 5, 37, 40, 54, 78, 79, 86, 112, 121, 138
  Curriculum, xxxi, 43
  Developmental, 25
  Institutional, 144
  Social, xxxi, 145
  Spontaneous, 41
  Transformative, 25, 31, 42, 75, 100
Chaos, 29, 32, 41, 43, 76, 85, 87, 99, 157
  And order, 2, 26, 30–31, 70
  Definition, 2
  King of, 1–2, 19, 151
  Theory, 19, 151, 173
Chaotic pedagogy, 157–159
Chaos and complexity theory, xii, 2, 25, 43, 75, 101, 121, 142, 157
Chaos and Complexity SIG, AERA, 142
Chebakova, Anastasia, xxviii, xxviii, 33, 55, 56, 61, 62, 124, 133
Chen, Yuting, 135
China, xix, xxvi, xxiii, 2, 12, 22, 46, 87, 96–101, 146, 150
  And John Dewey, 103–108
  Encounter with, xix, 99
  Teaching in, 102–103, 134, 135
Chomsky, Noam, 74, 75, 159
Clifford, Patricia, 7
Cochran-Smith, Marilyn, xxvii
Collaboration vs. competition, 9, 126, 130
Collaborative teaching, xvi, 57, 58, 126, 136, 139

Community, xvii, 19, 21, 33, 43, 50, 66, 92, 108, 129
  African American, 145
  Class, xxv, 47, 126, 130
  Dynamic, 158
  Holmes program, 17, 19, 44
  International, 130
  Local, 130, 140, 146
  Mutually flourishing, 144
  National, 130
  Of difference, 43, 82
  Of learners, 37, 140–144
  Of scholars, 143
  Organizer, 150
  School, xxx, 138–140
  Teacher learning, 60, 61
  Without consensus, 141
Competency-based education, 75
Complexity, ix, xviii, xxv, xxvi, 8, 21, 49, 54, 75–79, 89, 100, 120, 149, 151, 167, 169
  And simplicity, 84–86
  Definition of, 79
  In writing, xxv
  Of education, xviii, 86
  Of number system, 77
  Of teaching, xxvii, 58
  Theory, xxvi, 54, 74, 76, 79, 96, 105, 170
Conflict, xv, 18, 50, 104, 112, 116, 124, 126, 160–162
Confucianism, 104
Constructivism, 11, 36
Context, xxvii, 3, 13, 15, 36, 49, 61, 80, 81, 87, 154, 169
  Cultural, 21, 107, 120, 132
  Global, 37, 102
  Historical, xxix
  Of education, 12, 57
  Of teaching, 25, 51, 172
  Political, 87, 144
Control, xiii, 2, 48, 123, 138
  Centralized, 36, 50, 88
  Dissipative, 36, 41, 135, 150

INDEX

Emergent, xvi, 2, 126, 147
Male, 48
Self-organizing, 36
Teacher's, 36
Conversation, ix, xxix, 8, 11, 21, 45, 51, 64, 77, 95, 97, 115, 138, 147
  Class, 32, 36, 41, 66, 76, 83, 85, 93, 120, 136
  Complicated, ix, xviii, 10, 32
  Democratic, 124
  Embodied, xv
  Intercultural, 100, 134, 135
  Intellectual, 54, 59, 92, 142
  Interdisciplinary, 124
  Multilayered, xxix, 132, 135
Cornell University, xi, xxx, 5, 46
Cosmology, 21, 75, 167, 173
Cosmopolitanism, 98
Creative energy, 19, 39, 43, 94, 103, 148
Creativity, 2, 9, 11, 29, 36, 90, 94, 96, 102, 121, 149, 172
  And peace, 126
  As co-creative, 86, 126
  Teacher's, 26
Crossdisciplinary analysis, 161–162
Currere, 21
Curriculum, ix, xi, xviii, 2, 10, 16, 20–22, 34, 58, 63, 88, 108, 115, 120, 131, 138, 149, 163, 167, 171
  And chaos theory, 10
  Emergent, 122
  General education, xxxi, 146
  Pattern, 77
  Postmodern, 19, 20, 31, 63, 117
  Reform, 102, 108, 135
  Structure, 21, 30, 41, 71, 76, 119, 146
  Transformative, 20
  University, 41
Curriculum Camp, 92, 143, 156, 163
Curriculum studies, xvii, xxvi, xxix, 131, 143, 147, 152
  Internationalization of, xxii, 97, 98
  Worldliness of, 102

Curriculum Theory Project, xxxi, 10, 143, 144, 155
Curriculum theory, xxx, 2, 31, 57, 92, 143, 153–156, 164

D

Daigneault, Jacques, 155
Darling-Hammond, Linda, 17
Davies, John, 155
Davis, Brenda, xxxii, 38, 39, 106, 108
Davis, Brent, xxxii, 30, 36
Deconstruction, 21
Deliberation, 22, 133
Democracy, 19, 28, 66, 104, 107, 108, 130
Derrida, Jacques, 21, 39, 101, 135, 146
Dewey, John, xi, xxxi, 6, 8, 11, 13, 18, 19, 22, 28, 63, 66, 100, 103–108, 121, 130, 136, 139, 156, 164, 167, 172, 173
Dialogue, 1, 61, 125, 171
Difference, xi, xiv, 18, 20, 22, 26, 29, 40, 43, 48, 55, 60, 84, 108, 112, 119
  And peace, 116–118
  As complementary, 59
  As positive, 18, 39, 57, 62, 82, 83, 97, 98
  Connecting across, 22, 43, 122, 135, 140
  Cultural, 102, 134
  Intellectual, 26
  Learning from, 116
  Pedagogical, xviii
Disposition, 160–164, 174
Disequilibrium, xii, 25, 26, 29, 30, 160, 167, 172
Doll, Mary Aswell, xxviii, 4, 42, 48, 70, 90, 95, 115, 127, 138, 142, 145
Doubt, 1, 29, 171
Dynamics, xvii, 20, 21, 29, 46, 47, 61, 79, 108, 121, 136
  Class, 54, 103, 132, 137, 147
  Intellectual, 59
  Of participation, xvii, 149

E

Ecology, 71, 75, 96
Edgerton, Susan Huddleson, 19, 94
Efficiency, 34, 154
Ego, 121–122
Either/or, xii, xxi, 26, 28
Emergence, xiv, 30, 31, 36–39, 77, 79, 80, 85, 86, 91, 132
Emerson, Caryl, 122
Emerson, Ralph W., 168
Emotion, xiii, 10, 26, 30, 55, 90, 95, 122, 133
Enculturationist teaching, 163
Engagement, ix, xxii, 34, 50, 52–54, 62, 67, 117, 121, 132, 139, 157, 169
  Ethical, xxii, 98, 108, 147
  Intellectual, 132, 155
  International, 100, 147
  Learning, 49, 58, 65, 77, 96, 103
  Nonviolent, 118
  Pedagogical, 52–55, 67, 119
  Sustained, xiii, 50, 53, 54, 133, 148, 172
  With the other, xix, 99
Eppert, Claudia, 155
Ethics, xvii, 115, 144, 149
  Postmodern, 147
Experience, xiii, xviii, 5, 11, 29, 32, 47, 60, 65, 71, 85, 92, 101, 107, 108, 114, 122, 126, 136, 144
  Crafting an, 11, 14, 65, 94
  Educational, ix;
  Leadership, 41, 117, 138, 140
  Life xxvi, 19, 41, 66, 120, 150
  School, 4, 57
  Teaching, xxvi, 18, 48, 58, 118, 150
  Transformative, 123, 156
  Whole-being, 94, 133
  Working, 46, 116

F

Farber, Jerry, 49, 52, 53
Feminism, 136, 137

Feminist pedagogy, 137
Field experience, 64, 66
Finland, 102, 146
First among equals, 6, 8, 36, 173
5 C's, 20, 21
Fleener, Jayne, xiii, xxiii, 48, 54, 79, 155
Flow, xvi, 41, 54, 103, 105, 111, 132, 150
  Creative, 122
  Pedagogy in, 118–122
Foucault, Michel, xxiii, 9
4 R's, 10, 20
Fowler, Leah C, 55
Fractal, 10, 85, 87, 88, 96, 100, 101, 117, 154, 173, 174
  Entrance into, xxix, 75–78
Fraser, James W., 16
Freedom, xi, 4, 10, 52, 88, 154, 172
  Intellectual, xiii, 85, 142
  To explore, 3, 4, 17, 47, 112
Friday Friends, xv, 10, 140–142, 155
Friesen, Sharon, 7
Frobel, Friedrich, 80
Fu, Guopeng, xxxii, 103–108

G

Garrison, Jim, 100
Gender, xiv, xxi, 56, 60, 62, 87, 115, 137, 138, 144, 145
Genres approach, 159, 161
Gleick, James, 76
Globalization, 97
Goodson, Ivor F., xxvii
Great Books program, xxx, 7, 135
Griffin, David, 141
Guilt, xvi, 112, 114, 115, 122, 149

H

Hansen, David T., 98
Heidegger, Martin, 15, 18, 26

Hendry, Petra Munro, xiii, xxi, xxiii, xxviii, xxxii, 32, 48, 51, 58–61, 76, 77, 85, 90–92, 96, 119, 120, 136–138, 147, 148, 155
Heretical Catholic, 22, 148
Hermeneutics, 94, 173
Hershock, Peter D., 144
Hirsch, E. D., 82
History, xxv, 1, 4, 14, 17, 46, 52, 54, 80. 81, 94, 98, 104, 131, 154, 156, 157
Holmes Program, xxxi, 15–16
　Curriculum, 15–18
　LSU Elementary, 2, 14–19, 35, 41, 62, 64, 69, 144
Holt, John, 6, 7
hooks, bell, 137
hospitality, 19, 94, 97
Houston, Robert W., xxvii
Hoyt, Mei Wu, xxxii, 54, 56, 115, 135
Huebner, Dwayne, 22, 86, 123
Hull, Bill, 7
Humanism, 153, 154, 164
Humility, x, xii, xiv, 22, 71, 122, 147, 168
Humor, 14, 18, 20, 40, 53, 59, 95, 115

I

Imagination, 2, 50, 51, 69, 149, 168
　Mathematic, 87
Improvisation, xii, 10, 11, 18, 37, 95
Individuality, xiii, 13, 102
Inner child, 11
Interaction, xvii, xxv, 13, 14, 31, 36, 47, 58, 61, 77–79, 94, 125, 133, 138, 149
　Between theory and practice, 17, 64
　Class, 52, 57, 71, 86, 87, 103
　Dialogical, 81
　Dynamic, 17
　With children, 63
International Association for the Advancement of Curriculum Studies (IAACS), 97, 168

Interplay, xii, 13, 14, 17, 21, 75, 98, 102, 126, 150, 159
　Between passion and peace, 111
　Between presence and absence, 48
　Between the part and the whole, 76
　Between the self and the other, 100
　Between theory and practice, 19
Internship, 15, 17, 18, 67, 140
Intimate revolt, xii, 29

J

Japan, 104, 105, 107, 146, 147
Jacobs, Mary-Ellen, 155
Jardine, David, 7
Jenks, Charles, 141
Johns Hopkins University, xxxi, 4
Journey, xvi, xxv–xxvii, 2, 49, 55, 96, 97, 100, 113, 115, 122, 148–151, 170, 171, 173
Joy, xvi, xxvi, 15, 53, 60, 120, 169, 174
　And teaching, 93–96
　In students, 90–92
Jung, Carl, 2, 11, 49
Jung, Jung-Hoon, xxix, xxii, xxiii, xxix, xxxii, 12, 105, 125
Juxtaposition, 84, 158

K

Kane, Ruth, 5
Kauffman, Stuart, 22, 31, 79, 84, 88
Kieran, Tom, 26
Kirshner, David, vii, xxviii, xxxii, 20, 47, 61, 70, 73, 77, 140, 144, 153–164
Kliebard, Herbert, 145, 172
Knowing, 12, 50, 51, 93, 96, 119, 120, 148, 169
　Patterns of, 74
　Relational, 51, 75
Kohli, Wendy, xxviii, xxxii, 62, 121, 155
Korthagen, Fred, 64, 136

Kosnik, Clare, 17
Kristeva, Julia, xii, 29, 106, 113, 123
Kroll, Linda, 29, 30
Kuhn, Thomas, 159
Kundera, Milan, 149, 167

L

Labaree, David F., 16, 17
Labyrinth, 149–150
Latta, Margaret Macintyre, 15
Laughter, x, xv, 1, 4, 9, 14, 42, 91, 94, 95, 104–107, 140, 151, 167
Leadership, xxx, 2, 6, 19, 41–43, 92, 117, 130, 138–140, 144, 146
Learning, xxvii, 3, 5, 10, 26, 39, 40, 49, 55, 65, 79, 87, 93–96, 112, 116, 124
 About the other, 96
 As advertent, 162
 From, xix, 7, 17, 56, 96, 108
 Process, 11, 45, 48–50, 137, 160
 Theory, 159–164, 169
Leggo, Carl, 105
Letting go, 2, 117, 119, 147
Levine, Arthur, 16, 65
Li, Xin, xxvii
Life history, xxvii, 96, 169
 Pedagogical, xxvii, xxix, 2, 151
Lin, Jing, 116
Listening, xv, xxvii, 48, 52, 82, 92, 95, 119, 132, 133, 141, 171
 Indigenous ways of, 125–126
Literacy, 81–84, 86
Liu, Patricia, xxxii, 14–14, 125
Loss, xvi, 112–115, 122, 123, 127, 150, 168
 Silence over, 113
Loughran, John, 49, 64, 136, 150
Louisiana State University, xvi, xxv, xxxi, 1, 10, 46, 140, 143, 153, 168
Love, 2, 10, 19, 35–37, 91, 114, 168, 169

M

Mandelbrot Set, 76
Mann, Steve, xxi, xxiii, xxxi, 4, 28, 29, 35, 112, 145, 149
Marsden, Rasunah, xxxii, 83, 84, 105, 125
Marsh, Colin J., 86
Martusewicz, Rebecca R., 30
Mayes, Clifford, 2, 49
McGuire, Michael, 101
McCarthy, Cameron, 155
Metanarrative, 146
Miller, James, 9
Miller, Janet L., 84, 92, 102, 137, 141, 152
Mitchell, Roland, 155
Merleau-Monty, Maurice, 83
Mobilization, 120
Modeling, 19, 57, 61, 63, 124, 150, 161, 171
Montessori, Maria, 70
Morson, Gary Saul, 122
Multiple perspectives, 33, 77, 131
Mutuality, xix, 58, 99, 100, 108
Mysterium tremendum, 21

N

National Council for Accreditation of Teacher Education, 160
Nationalism, 97
Negotiation, 10, 18, 43, 55, 67, 99, 117, 118, 144, 149
Noddings, Nel, 64
Nonattachment, 119
Non-belonging, 117
Nonlinear teaching, xxix, 2, 31, 39, 63, 64, 74, 77–84, 102
Nonviolence, xv–xvi, 119, 126

O

Old wise man, 2
Olson, Nadine, 56

INDEX

Oswego–Sotus Structural Arithmetic Project, 73, 75
Oxford, Rebecca, 121

P

Palmer, Parker, 22
Pang, Jeong Suk, 64, 87
Park Elementary School, xxx, 6
Parade child, 1, 2, 151, 171
Participation, x, xvi–xvii, 22, 32, 43, 126, 129–151, 155, 160
  Democratic, xvii, 132, 133, 138, 145
  In class, 87, 130
  In community, 43, 130, 163, 172
  Parental, 139
  Philosophy of, 58, 130–134
Passion, xv–xvi, 10, 22, 73, 89–108, 111, 149, 174
  And peace, 111, 126
  And play, 89, 94
  For life, 2, 11, 91
  Intellectual, 63, 144
Patience, xvi, 53, 56, 90, 118, 122, 141
Pattern, xiv–xv, 10, 11, 17, 20, 22, 30, 69–88, 97, 113, 149, 151, 157, 171–173
  Complex, 76, 78–80
  Ecological, 71
  Emergent, 31, 79, 81
  Mathematical, 70–73, 78, 87, 89
  Nonlinear, xiv, 71, 74, 77, 99
  Of subject matter, 70–74
  Within pattern, 76
Peace, xv–xvi, 22, 111–126, 149, 151, 171, 172
  As working with difficulty, 116
  Definition of, 111
  Dynamic, 111, 126
  Inner, xv, 111, 115, 121
Pedagogical authority, 2, 86, 118, 138
Pedagogical bonding, xiii, 46
Pedagogical companionship, xiii, 50–53, 172
Pedagogical conditions, 79, 93, 134

Pedagogical creed, 118–119, 173
Pedagogical design, 85
Pedagogical distance, 6
Pedagogical effect, 32, 33, 46, 51, 111, 115, 122–126
Pedagogical relationship, xv, 6, 33, 35, 48, 50, 115, 122–124, 148
Perturbation, xii–xiii, 18, 22, 25–43, 47, 79, 118, 149, 151, 171, 172
  Conditions for, 33–39
  Limit of, 39–41
  Positive role of, 26, 28, 30, 99
Piaget, Jean, xi, xii, xxi, 11, 25, 26, 29, 30, 32, 63, 70, 74, 75, 121, 136, 141, 145, 160, 167, 172
Pinar, William F., ix–xix, xxv, xxvii–xxxii, 10, 52, 59, 65, 71, 88, 90, 94, 97, 98, 102, 120–122, 135, 142–145, 147, 152, 154
Play, x–xii, xxix, 1–22, 47, 69, 74, 78, 88, 94, 96, 111, 122, 149–151, 171
  And work, 3
  Complex, xi
  Intellectual, 53, 89, 11–15, 95, 142, 163
  Passionate, 94
  Postmodern, 20, 157
  Serendipitous, 20
  Spiritual, 3, 95
  With difference, xvii, 14, 18, 84, 136
  With difficulty, 90
  With limits, xi, 9–11, 111
  With patterns, 69, 71–74
  With relations, 15, 18–19, 111, 144
  With subjects, xi, 3, 6–9, 61, 70, 139
Politics, xvii, xxv, 17, 144–147, 149, 154
  As power struggle, 146
  Identity, 144
  Participatory, 147
Porche-Frilot, Donna, 56
Postmodern theory, 64
Postmodernism, xxi, xxvi, 46, 52, 75, 98, 105, 117, 118, 131, 154, 157, 164, 173
Postmodernist, xxvi, 52
  Optimistic, 117

Paternalistic, 52, 115
Post-structuralism, 117, 168
Potentiality, 87, 90, 93, 119
Power, 35, 130, 145–147
Pragmatism, 46, 105, 136, 137, 170
Pratt, Sarah Smitherman, 63, 77, 86, 93
Praxis, 155
Presence, xiii–xiv, xxix, 22, 45–67, 92, 99, 109, 115, 138, 149, 151, 167, 168, 171, 172
  And absence, 47–49, 55–57, 115
  Intellectual, 94
  Mutual, 60, 62–67
  Personal, 64
  To the self, 49
Prigogine, Ilya, xi, xxi, 2, 25, 26, 29–31, 78, 79, 121, 141
Privilege, xix, 56, 59, 99, 115, 122
Professionalism, x, xvii
Psychology, 14, 153, 156, 159–161
Puritan tradition, 113–114, 132

Q

Questioning, xix, 21, 26, 39, 53, 66, 100, 144, 149, 168
Quinn, Molly, vii, xxxii, 21, 116, 121, 167–17

R

Race, xiv, 56, 95, 137, 138, 144, 145
Ramus, Peter, 131
Reciprocity, xiii, xx, 15
Reconceptualism, 152, 154
Recursion, 10, 52, 74, 77, 107
  Definition of, 20
Reference culture, 161, 162, 164
Relation, xii, 3, 7, 10, 15, 19, 49–53, 64, 77, 94, 137, 153, 158, 159, 169, 170, 172
  Cultural, 87, 88
  Definition of, 20, 87
  Social, xv, 9, 18, 111
Relational dynamic, xvii, 147
Relationality, xiii, xv, 45, 50, 88, 172
  Complexity of, 49
  Ethical, 147, 148
  Organic, 52
Responsibility, xvii, 36, 50, 146
  Doubled dimensions of, 149
  Ethical, 130, 149
  Political, 130, 159
  Social, 145
  Spiritual, 130, 149
Retreat and create, 142
Reynolds, Sherrie, 36, 73, 78
Richness, xxvii, 10, 84, 85, 169
  Definition of, 20
Rigor, 10, 96, 120, 164
  Definition of, 20
Robertson, Emily, 132, 133
Roman, Leslie, 155
Rorty, Richard, 146, 173
Rousseau, Jean-Jacques, 80
Roy, Kaustuv, 155
Russell, Tom, 64, 136
Russia, 33, 55, 56, 102, 124, 146

S

St. Julien, John, 10, 16, 54
Salaam, Tayari Kwa, xxviii, xxxii, 33, 56, 57
Samara, Dennis, 30
School deform, 65
Schubert, William, xxvii
School board, 43, 146
Science, xii, 4, 21, 88, 108, 117, 149, 156, 157
Scientific management, 34, 131
Self, xii, xviii, 21, 22, 49, 61, 83, 90, 100, 106, 111
  And the other, 100, 126, 135
  And the world, xvi, 122, 126
  Ecology of, 96
  In system, xviii, 96, 121

INDEX

Notion of, xxvi
Teaching, xviii, 96
Sense of, 21, 96, 106, 116
Self-organization, xviii, 31, 33–36, 75, 76, 79, 88, 91, 99, 136, 148, 151
  Conditions for, 84, 86
  Definition, 78
Self-education, xix, 99
Serres, Michel, xx, 15, 18, 22, 66, 130, 146, 147
Sharing, xiii, xv, 10, 45, 50–53, 90–95, 123, 124, 126, 147, 169, 173, 174
Shen, Heyong, 150
Sikes, Pat, xxvii
Sikula, John, xxvii
Situated cognition theory, 159
Situation, x, 22, 26, 36, 39, 42, 65, 86, 102, 115, 121, 144, 145, 170, 172
Skinner, B. F., 141, 159, 160, 162
Social justice, 30, 137, 145, 146, 148
Spirit, xii, 4, 14, 20–22, 56, 57, 80, 88, 91, 125, 170
  Dynamic, 89
  Joyful, xv, 89, 93
  Of Christ, 22
  Of creativity, 34, 61, 172
  Of subject matter, 71, 94
  Playful, 2, 4, 5, 6, 25, 94, 143
  Postmodern, 117
  Wholeness of, 100, 148
Spirituality, xii, xvii, 22, 85, 88, 122, 133, 138, 144, 149
Spontaneity, xiv, 3, 39, 52, 58, 78, 150
Stanley, Bill, 155
Stengel, Barbara S., 50, 51
Stengers, Isabelle, 29, 31, 78, 79
Story, 8, 21, 88, 100, 105–107, 146
Story-sharing, xxvii, xxviii, xxxii, 169, 173
Strange attractor, 79, 81, 157
Structure, 21, 30, 37, 42, 58, 75, 85, 157, 160, 162
  And surprise, 76, 81
  Definitions of, 30
  Dissipative, 29, 30, 36, 148

Playing with, 9
Subject, 6, 71–73
Study, ix, xx, 25, 83, 84, 155
Subjective formation, xxvi
Subjectivity, xviii, xxvi, 96, 121, 122, 145
  Dissipative, 170
Suffering, xv, xvi, 122, 150
SUNY Oswego, xxi, 32, 41, 73, 135, 141
Syllabus, 7, 8, 10, 11, 37, 40, 60, 76, 85, 93, 146
Symbiosis, 149

T

Taoist dynamics, 29, 54
Taubman, Peter, xxi, 6, 10
Teacher education, xxvi–xxvii, xxix, 14, 26, 30, 57, 74, 81, 120, 126, 136, 140, 150, 169 critiques of, 15–16, 60, 64–65, 93, 95
  History of, 16, 34
  Instrumental approach to, 50
  Pedagogy of, xxvii, 36, 49, 67, 151
  Programs, 15–19, 140
  Reform, 65
  Standards movement in, 34, 119
Teacher educator, xxv–xxvii, 2–3, 29, 36, 61–64, 93, 126, 141, 169
  And play, 9–19
  Responsibility of, 131
  Women, 138, 150
Teaching and learning, xiii, xv, 7, 11, 26, 39, 50, 77, 85, 93, 95, 121, 126, 172
Team teaching, xxxi, 23, 32, 40, 57–62, 85, 96, 124, 135–138, 142, 147, 150
Temporality, 79, 86–88, 91, 118
Tension, 13, 16, 18, 21, 41, 67, 116, 124, 126, 135, 149, 151, 154, 160
  Between chaos and order, 70, 79
  Between play and passion, 94, 96
  Creative, xi, 21
  Right amount of, xiii, 25, 29, 32–33, 39

Theory and practice, xi, 16, 22, 62–67, 145, 150, 151
Third instruction, 148
Third space, 94, 96
3 S's, 20, 21, 107, 149
Time, x, 3, 10, 79, 86, 91, 132, 138, 156
   Recursive, xiv, 52
   Role of, 33–35, 87
   Postmodern, 87
Todd, Sharon, 96
Train, Pete, 81–83, 106
Transformation, xiii, xvii, 2, 49, 75, 77, 96, 106, 121, 136, 138, 145, 149–150, 157, 163, 170
   Cultural, 88, 135
   Curriculum, 11, 20
   Recursive, xii, 29
   System's, xii, 25, 36, 77
   Through team teaching, 136
Translation, 62, 97 134–135
Transcendence, xv
Trauma, 19, 112, 122, 127
Triche, Stephen, xxviii, xxi, xxxii, 45, 52, 53, 54, 55, 90, 117, 118, 123, 124, 130, 148
Tröhler, Daniel, 12, 14, 37, 38, 80
Trueit, Donna, xi, xiii, xiv, xvii, xviii, xx, xxii, xxiii, xxv, xxviii, xxix, xxxi, xxxii, 2, 11–14, 23, 26, 29, 33, 37, 40, 41, 48, 52–56, 58, 59, 61–63, 79, 87, 93, 95–97, 100–103, 109, 113–115, 120, 121, 124, 125, 129, 133, 138, 143, 170, 175
Tyler rationale, 20, 131
Turbulence, 18, 43, 112, 118, 159

U

Uncertainty, xii, 19, 26, 35, 43, 79, 148, 168
Unity between teacher and teaching, 45
University of British Columbia, xxviii, xxxi, 12, 89, 102, 103
University of Redlands, xxxi, 141

University of Victoria, xxxi, xxviii, 40, 55, 58, 89, 124, 129, 134

V

Valley School, xxx, 9, 70, 138–141
Vygotsky, Lev, 160

W

Walkerdine, Valerie, 155
Walter, Karya, 101
Waltman, Shreyl, xxviii, xxxii, 11, 18, 35, 65, 66, 67
Wasko-Flood, Sandra, 126
Whitehead, Alfred North, xi, 72, 82, 136, 141, 171
Whitson, Tony, 154, 155
Willis, George, 86
Win-or-lose mentality, 9, 126, 147
Wisdom, 19, 26, 27, 49, 66, 101, 148, 168, 174
Wonder, x, 71

Y

Yu, Jie, xxxii, 56, 76, 85, 101

Z

Zero, 73, 78